Rebirth of American Industry

A Study of Lean Management

By
William H. Waddell
Norman Bodek

PCS Inc.
PCS Press
809 S.E. 73rd Avenue
Vancouver, WA 98664

bodek@pcspress.com
http://www.pcspress.com

Printed in the United States of America

Printing number
1 2 3 4 5 6 7 8 9 10

The cover is a stereogram designed by Gene Levine

Library of Congress Cataloging-in-publication Data

Waddell, William H. and Bodek, Norman.
Rebirth of American Industry – A Study of Lean Management

Cover design – Gene Levine
p. cm.
Includes index.

ISBN 0-9712436-3-8
1. Production management. 2. Industrial Management. 3. Organizational change. 4. Organizational behavior 5. Manufacturing processes.

The Phoenix

The Cover: The picture on the cover is a stereogram: Hidden is a three dimensional picture (3-D) by the artist Gene Levine. If you relax your eyes, focus six inches beyond the top of the book, or look at the blank space above the picture, or hold the cover to your nose and slowly move the page away, you should see this wonderful image emerge.

The Phoenix is an ancient symbol, representing rebirth. It is a bird of great beauty. Wherever it is found, the Phoenix is associated with resurrection, immortality, triumph over adversity - a new life which rises out of the ashes. For America, it is symbolic of the 21st century resurgence that is coming in manufacturing from the ashes of abandoned plants and decimated towns through the adoption of lean manufacturing management principles.

BILL WADDELL'S INTRODUCTION & ACKNOWLEDGEMENTS

After the manuscript for this book was written, but before it went to publication, the $28 billion Delphi Corporation filed for bankruptcy protection. Delphi has long been considered a U.S. automotive industry lean manufacturing success story, and has been awarded 24 Shingo Prizes for lean manufacturing excellence. Lean manufacturing in some of the Delphi plants, however, was not translated into sufficient financial success for the company to succeed. Inventory turns and the overall costs of Delphi were, at the time of the bankruptcy filing, about the same as they were when the company was split from General Motors.

This very high visibility lean manufacturing failure has become the subject of quite a bit of discussion within the manufacturing community, much of it revolving around the fact that the Delphi case is not unique. Ford has also won a number of Shingo Prizes, but has not been able to make the manufacturing changes necessary to bring it closer to the performance levels of Toyota. The failure of lean manufacturing to succeed is not unique to the automotive industry. There are a number of manufacturers in other sectors, particularly in defense, who have become lean in some of their factories – and some have even won Shingo Prizes – but have generally failed to bring those visible plant results to their bottom lines. The automotive industry is simply the biggest and most visible sector of manufacturing and subsequently the focus of the popular press.

Delphi management's reaction to its clear failure to generate lean manufacturing financial success is also typical. Rather than delve into why they failed to achieve Toyota-like cost reductions and inventory turnover rates, they are blaming their bankruptcy on the 'legacy costs' – pension and health care commitments – they are obligated to pay based on contracts GM signed with the UAW long ago. They also blame the failure on the hourly wage rates they are contractually bound to pay their union employees, and they are trying either through direct negotiation or through the force of the bankruptcy court to cut wage rates by two-thirds. They also blame energy and material costs and are bantering about plans to close more U.S. plants and move more production work to lower cost countries.

The aforementioned scenario has played itself out over and over again in the United States within the last twenty or thirty years. It was around that time that Toyota began to take over huge chunks of the market shares from big American companies through its high powered, lean manufacturing strategies. Since then, one American company after another has found itself unable to go head to head with lean Asian competitors. Many of these companies found failure in their own lean manufacturing initiatives, and often pursued strategies of outsourcing work from the U.S. to low labor cost countries.

The objective of this book is to explore the near unanimous failure of the big, publicly held American manufacturers to become Toyota type lean manufacturers. The fault seems to lie in the basic financial assumptions that form the core of manufacturing management. The evolution of lean manufacturing through Henry Ford and the Toyoda family is outlined and compared to the origination and evolution of traditional American management from Alfred Sloan and General Motors through today.

Please keep in mind while reading *The Rebirth of American Industry* that our purpose was to write a management book, rather than a history book. While the historical bases for lean and traditional manufacturing management are explored, this is hardly a comprehensive manufacturing history. There is little, if any, new historical ground plowed in this book. We have simply compiled well known facts and sifted out only those that relate to the evolution and underpinnings of two contrasting approaches to manufacturing management. Key characters such as Henry Ford, Alfred Sloan and the creators of Toyota were all extraordinary, complex people. Dozens of excellent books have been written by and about these people and their ideas; most of which provide a much more precise and comprehensive understanding of them. We have taken only the key aspects of their lives and theories necessary to understand the effect they had on the evolution of manufacturing management, and the impact their theories made on lean manufacturing. Even in this narrow context, we are certainly presenting an insufficient picture of these men and the events of their times. We strongly suggest that the interested reader find and read the biographies and business books listed in the reference section for a fuller, fairer understanding of the players in the development of manufacturing management.

While we identify Pierre DuPont's Return On Investment (ROI) formula as the critical factor that prevents American manufacturing from becoming as lean as the Japanese, this book is not intended to be a comprehensive analysis of ROI theory either. The library shelves of academia are groaning under the weight of hundreds of expert analyses of ROI theory and the various modifications to it that have been proposed and occasionally adopted since DuPont and Sloan's time. Our concern is with a very narrow aspect of ROI theory, particularly the relationship between cash and inventory and how manufacturing management has applied the ROI based relationship between the two. We are also concerned with the Return On Sales element of the formula. For a complete understanding of ROI we recommend any of the countless sources available that give the subject a much more complete and accurate discussion than we have.

A pure accounting theorist will also note that under the broad heading of ROI we also discuss other aspects of manufacturing accounting that, technically, are not part of ROI. For instance, the GAAP principles regarding balance sheet items are not an element of ROI. Instead, ROI builds on those principles. Our purpose is to explain the overall implications of two fundamentally different approaches to manufacturing economics for the benefit of non-financial, manufacturing people. We hope the informed reader will forgive our occasional over-simplification.

Finally, there are a number of people who have had a profound impact on this book and the development of the ideas it contains. First and foremost is my co-author, Norm Bodek, who is actually the unrecognized co-author of just about every book on lean manufacturing published in the United States in the last 25 years. Norm almost single handedly brought the Toyota Production System to America through more than 60 trips to Japan, and through his tireless efforts to cultivate and grow relationships with the often reluctant Japanese experts. Any lean 'expert' who sits down with pen in hand to share his knowledge owes a great deal of that knowledge to Norm's commitment to lean manufacturing.

Credit also goes to Professor Bob Amsden of the University of Dayton. He has forsaken the more lucrative path to academic success that entails publishing the results of profound thinking and research in academic journals, for the lesser paid path of devoting a lifetime to churning out hundreds of students armed with his passion and practical knowledge of the tools to make American manufacturing quality the best that it can be. His curiosity and encouragement in the early days of developing the concepts in this book meant more than he will know.

Thanks to the many good people at Northern State University for their support. This book would not have happened without access to your knowledge, facilities, bottomless coffee pots and deep wells of kindness.

I also owe a large debt to two great teachers who went about life disguised as manufacturing managers at the Copeland Corporation – Dean Ruwe and Paul Maranka. They were great managers because they passionately pursued new ideas in manufacturing and demanded that everyone around them to do the same.

My deepest thanks go to Joan Schroder, Christopher Villegas, and Alfredo Caudilla. They are, in order, the best Purchasing, Operations and Materials Management professionals I have ever worked with. It was a blessing to have had them working for me for the brief period we had together. This boss is eternally grateful for the opportunity to learn from such outstanding people.

I owe my greatest debt to Bill Waddell, senior. He has forgotten more manufacturing information than I will ever know, and has instilled a passion for manufacturing in me from such an early time. I cannot recall there ever having been a day in my life when I wanted to do anything other than follow in his footsteps and be a manufacturing manager, and a father. He taught me that manufacturing can be much more than simply a way to earn a living. When done well, manufacturing management is a means of giving to everyone around you. Thanks, Dad, for the mentoring, editing, and mostly for the unconditional support.

Finally, there is Steve Baldwin. Only a true giant can make stepping over life's highest hurdles look so effortless. He lives cheerfully and gracefully, showing all around him how extraordinarily ordinary men can live. His example has enabled me to keep any problems in my life in perspective. I am truly honored to have known him for so long.

Bill Waddell

NORMAN BODEK'S INTRODUCTION & ACKNOWLEDGEMENTS

There is a certain amount of sadness when one reads of the bankruptcy of Delphi Corporation, and the losses and downsizing of General Motors (GM) and Ford. GM and Ford were the world leaders in automobile manufacturing and two of the richest companies in the world. The question is, what happened? Could it have been prevented? And have they, and we, learned the very important lesson of not repeating the mistakes of the past? The very purpose of this book is to examine the principle things that went wrong, and to give modern managers specific guidelines to think about in order to be internationally competitive.

In my own personal journey studying Toyota and the other leading Japanese manufacturers these past twenty-five years, I have been continually amazed at what I have discovered and am continually puzzled as to why GM and Ford have not taken the same steps I did. To learn from the best and then to forcefully apply the new methodologies into the company is a necessary step for improvement.

On my first study mission to Japan in February 1981, I visited the American Embassy in Tokyo and met with the information officer. His job was to study the best Japanese technologies and to have that information translated to English to help American companies stay abreast of what was happening in Japan. I was furious at him for wasting millions of dollars on research and not discovering what Toyota was doing to go from producers of "junk" to world class manufacturing.

Somehow, I was surely blessed to have met Dr. Shigeo Shingo, Mr. Taiichi Ohno, Dr. Ryuji Fukuda, Seiichi Nakajima, Dr. Yoji Akao, Hiroyuki Hirano, Shigehiro Nakamura, Bunji Tozawa, Iwao Kobayashi, Kenichi Sekine and others who were willing to share their information with Americans and allowed me to publish their Japanese books in English.

From my frequent visits to Japan, 65 as of this date, it became obvious to me that the Toyota Production System (TPS) was the most important and the most valuable process to study. When I first met Mr. Taiichi Ohno, vice-president of manufacturing at Toyota, I asked him to let me have information in writing about TPS. He said, "Norman, we don't have things written down, for it is always changing." I felt that he was just reluctant to share the information that was making Toyota so successful, but I was perseverant. I wouldn't stop searching for information to share with companies in the West. Magically I found Dr. Shingo, co-creator with Ohno of TPS, and he graciously allowed me to publish all of his books in English. After a few years, Mr. Ohno also gave me permission to publish his books in English.

Why didn't GM and Ford do as I had done and learn from these great works? Why wouldn't they locate, translate and publish everything available on Toyota? It is still a mystery to me. In 1984, Toyota decided to open a joint venture plant with GM, NUMMI, in order to share their production system. Following this venture, virtually all of Toyota's secrets were now available to GM. Why didn't GM study carefully the JIT/LEAN concepts and apply them? I owned Productivity Press and published most of what Toyota was doing. This knowledge was public, why did GM choose not to use such vital and public material?

I am sure many companies then and now have tried to study Toyota's production system and become Lean. But, why aren't they Lean? In retrospect, Americans weren't looking at the power and simplicity of TPS/JIT/LEAN. For some strange and unknown reason, our pride or maybe our feeling that we did things better than others, restricted us from "copying" something that didn't originate in the U.S. At the time, GM was the best in the world. One could speculate that they thought everyone should be copying them, not the other way around.

For the first twenty years of my studying manufacturing techniques, I almost solely looked at the TPS system; all of the marvelous tools and techniques (SMED, Poka-Yoke, one-piece flow, TPM, Hoshin Kanri, Kanban, and others)-without realizing that Toyota's success rested on two pillars, not just one. "The elimination of all non-value adding wastes," was one, while "Respect for People" was the other. Yes, GM and Ford studied the first pillar, but without engaging and understanding the second; they were doomed to failure.

Toyota develops people. To my knowledge, Toyota has never laid off a single worker since 1949. They train people, they educate them, they build their skills and capabilities, and they engage them (continually asking them and empowering them to come up with small creative ideas to make their work easier and more interesting). One of the earlier books I published and had translated from the Japanese was, *40 Years, 20 Million Ideas - The Toyota Suggestion System*, by Yuzo Yasuda. Here was a very simple but powerful system and probably the greatest key to Toyota's success (which was almost totally ignored in the West). Ironically, much like TPS, it originated with Henry Ford. The suggestion system, now called Quick and Easy Kaizen, was created in 1898 by Kodak.

In my own little way, I have tried to be like Paul Revere but instead of shouting "The British are coming!" I shouted "The Japanese are already here!" Numerous times I have tried to reach senior managers at Delphi, GM and Ford but no one would call me back. Unfortunately, I wasn't smart enough to get through their layers of secretaries. And even though I have published hundreds of books on Toyota, quality, productivity, etc., never has a GM executive ever picked up the telephone to have a dialog with me.

Here in this book, primarily through the efforts of Bill Waddell, we offer you a deeper understanding of the current "mess" self-created by Delphi, GM, Ford and others. Primarily, the problems faced by these giants came from following the wrong production system, not engaging employees to their full potential, and not having clear "measures" to drive continuous improvement.

How could GM make $3.6 billion in 2004, celebrate with huge bonuses to top managers, and six months later have their bonds declared "junk"? We hope that as you read this book, in addition to having lots of fun, you see a clearer path for a rebirth of American Industry – learning from past mistakes.

This book is dedicated to all those committed to working for a rebirth of American Industry. Yes, we are challenged to meet international competition, probably more so today than ever before. Hopefully, our book will "rattle" you out of your complacency, and inspire you to look for and find new ways to meet these challenges. I like the phrase, "When one door closes another one opens." It is up to us to take a deep breath, and look around for those new ideas and those open doors. They are always there.

Years back, a commonly used phrase was "Work Smarter, Not Harder." I never really liked that phrase at all. Looking inside factories you could see people never being asked to use their brains. How could they work smarter if they were never asked for their ideas?

I thank all of my past teachers, even those who gave me a rough time, as most did, for they gave me the determination to stick with things. It just never comes easy for us. But, I believe miracles do happen if you just continue to do your very best and persevere. I also want to thank T. V. Suresh of Tao Consulting for his help on editing the book and to Brian Maskell, etc. .... for their kind remarks.

Thank you very much.

Norman Bodek

Foreword

In *Rebirth of American Industry*, Bill Waddell and Norman Bodek provide a long overdue revision to the standard historical interpretation of the financial control system that DuPont brought to General Motors Corporation to save its huge investment in the floundering young auto company in the early 1920s. That system, installed at GM by senior DuPont executives Donaldson Brown and Albert Bradley and a young GM official named Alfred Sloan, is credited by historians with saving GM from financial ruin in the '20s and with carrying GM to unprecedented financial success in the next half century. Moreover, the Brown/Bradley/Sloan financial control system of "centralized control with decentralized responsibility" has been touted by business gurus such as Peter Drucker and Tom Peters and by leading graduate business schools as the gold standard of good management in American business from the 1950s to the present day.

From the vantage point of 2005, such praise for the "Sloan model" of management seems woefully misplaced in light of the currently unfolding demise of General Motors and countless other multidivisional corporate giants in the Fortune 1000 league. It now seems likely that "lean management," a quite different management model that Toyota resurrected after World War II, could replace the Sloan model as the management system of the future. But lean management's capture of the spotlight is not inevitable.

Indeed, so long as top managers remain committed to the manage-by-results "Sloan culture," Waddell and Bodek believe that companies have no hope of adopting the "lean culture" that permeates Toyota's remarkably successful system.

Ironically, as Waddell and Bodek point out, the famous American manufacturer known as the Ford Motor Company practiced lean management principles with fabulous success from about 1914 until the mid-1920s, at the very time that GM, with DuPont assistance, was creating the Sloan system and long before the Toyoda family ever thought of making automobiles. Historians have now established fairly well that Henry Ford and his people ran their Highland Park plant in Detroit according to the same principles that would later shape Toyota's operations – quality, flow, synchronization, and a belief that shorter time from start to finish (cycle time/lead time) is the pathway to lower cost. These principles would later appear only in part at Ford's more famous River Rouge plant and they would be entirely absent from all Ford operations by the 1950s. Such principles would never appear at all in GM operations, certainly not after top management began to guide the company according to the principles implicit in the Sloan model.

The central concern of this book is to outline the core principles in the Sloan management model that GM adopted after 1920 and to show how adhering to those principles makes it virtually impossible for managers to understand and adopt the principles inherent in Toyota's so-called "lean" operating system. The book deals less with the details of lean management, a subject that is covered extensively in the vast and growing literature on the Toyota Production System (TPS), than it does with the much less extensively covered details of the Sloan management system. The gist of the Sloan system is to rationalize and reinforce batch and queue operations, replete with massive inventories and extensive delays. The gist of TPS, of course, is to encourage continuous flow and reduction of overall lead times.

I feel at home commenting on the details of the Sloan system because I contributed in the 1970s and 1980s to the literature extolling what were seen then as virtues of that system. Later I came to understand the Sloan system's flaws, after a long period in the 1990s observing and studying TPS first-hand in Toyota factories. In writing recently about the need to replace batch and queue with lean I have commented briefly on the flaws in the Sloan model, but I never spent time discussing those flaws in detail. The flaws had become so self-evident to me that it seemed redundant to dwell on them. I now think I was mistaken to take that stance. I have come to believe that one reason many companies find it so hard to adopt Toyota's system is their failure to see how the financial thinking implicit in the Sloan system blocks out the thinking implicit in TPS. By helping to remove that block, Waddell and Bodek's book provides a badly needed complement to the literature on lean management.

It is instructive to look back, as Waddell and Bodek do, at how satisfied GM was with its methods for so many decades after the 1920s. By the 1950s, most large-scale businesses and almost all graduate MBA business programs manifested the same contentment with the Sloan-style of management by results. Everybody who was anybody seemed to be using that system of financial management, and all made lots of money. Unrecognized from the 50s through the 70s (and even later in many cases) was the fact that most American businesses had no competitors outside the U.S. The American economy was living in a dream world, where wages rose regularly each year and prices were automatically raised to match rising costs. Almost nobody was paying attention to the dead-weight burden of overhead costs that was sapping the strength of American businesses from within. The Sloan system unconsciously disguised that burden, and ultimately made matters worse, by encouraging the idea that companies could always cover those costs by raising prices, cutting out direct labor, and "selling" overhead to the balance sheet through increased product inventories. Meanwhile, profits, executive bonuses, and workers' wages kept rising.

The book also does well at showing how the "magic" of MRP software after 1960 further disguised the dead-weight impact of overhead cost and the financial management systems that stimulated its growth. The authors properly describe MRP as a sophisticated effort to rationalize the damage to sound operations caused by following the Sloan model of financial management. Citing a story told by Taiichi Ohno, they ask "would a manufacturing company that was operating on a philosophy of lead times of zero and lot sizes of one have any use whatsoever for an MRP system?" The answer, of course, is no. Indeed, Toyota does not use MRP to schedule the release of work to the floor or to route work through processes.

Another point, equally important, in my opinion, is the fact that Toyota does not use financial control systems to motivate, evaluate and control work on the shop floor. As evidenced by their avoidance of both MRP and management accounting control systems, Toyota does not use external information to control shop floor operations. The Toyota Production System is its own guide. All the information needed to guide and control operations is in the work – the material flowing through the processes and the standards for doing work contain all the information needed to get the job done correctly and on time. I wish the book stressed that point more. This is an important matter because there is today a growing interest in what many people call "lean accounting." Unfortunately many advocates of lean accounting are telling lean companies to adopt financial control systems that one never sees in a Toyota plant, but that are well known to companies that follow the Sloan-style model of financial management. In my opinion, this advice can be fatal to lean management. It should be rejected by any company that is serious about following Toyota's lead.

Waddell and Bodek call for restoring American industry by replacing the Sloan model of financial management with lean management. Rightly, they refrain from describing in any detail what form this restoration might take. That issue is the next frontier of research and practice concerning lean management. Some people, I am afraid, see lean as a pathway to restoring the large global manufacturing giants the United States economy has been famous for in the past half century.

Overlooked in this picture are the unfortunate environmental consequences of building such global production systems. The cheap fossil fuel energy sources that have always supported such production operations cannot be taken for granted any longer. One proposal that has great merit is that of rebuilding our economy around smaller scale, locally-focused organizations that provide just as high a standard a living as people now enjoy, but with far less energy and resource consumption. Helping to create the sustainable local living economy may be the most exciting frontier yet for architects of lean operations. Time will tell. In the meantime, Waddell's and Bodek's book helps mark the way by making us more mindful than ever of the pitfalls that lie waiting if we continue to follow the precepts of Sloan-style financial management.

- H. Thomas Johnson, Lake Oswego, Oregon, December 2005

Dr. Johnson is the author of *Relevance Lost: The Rise and Fall of Management Accounting, Relevance Regained, Profit Beyond Measure: Extraordinary Results through Attention to Work and People*

TABLE OF CONTENTS

CHAPTER ONE

The Magnitude of the Problem

"A man will spend thirty years in building up a mountain range of facts with the intent to prove a certain theory; then he is so happy in his achievement that as a rule he overlooks the main chief fact of all- that his accumulation proves an entirely different thing."

-*The Bee Essay* - **Mark Twain**

Manufacturing is important. Manufacturing has been the essence of our wealth and well-being. It should not be allowed to disappear from our shores.

Regardless of how much hopeful talk there is of the service economy or the transition from an industrial age to an information age, manufacturing is still the heart of any economy and the heart of America. Information and service industries have grown enormously but they do not grow our food, provide the tools used to build houses, or create the products we need to survive. Only manufacturing, mining and farming do that. Information and services produce wealth to the extent that they enhance manufacturing, mining and agriculture, but for the most part, all they do is simply shuffle the money around from one pocket to another.

As one great manufacturing company after another drives to outsource manufacturing overseas, we should have great concern. While American manufacturing companies are building plants in China because they believe they cannot compete with American labor and overhead costs, Japanese firms are building plants in the U.S. and competing quite well.

How can Toyota and the other Japanese companies do well in North America while American managers feel that their only option is to send work offshore?

Every stakeholder in the once great American manufacturers is suffering. The companies themselves are struggling for profitability, and some are gone completely. The market value of General Motors is dropping to a point near that of Harley-Davidson, which means investors have lost quite a bit of very real money. Long term employees have been laid off permanently, and the unions that represented them are almost gone. Few have gained from the course of American manufacturing over the decades since we first encountered foreign competition in the 1970s.

Understanding why American manufacturing has struggled, and for the most part failed, to become more competitive by becoming "lean" is not only important for the future of manufacturing companies themselves and for all of their various stakeholders, it is important to the country as a whole. Outsourcing to cut costs may seem to help stockholders in the short term. It may seem to keep costs down for consumers. In the long run, however, an American company that becomes successful by transforming itself into an American company in name only fails everyone. It is no longer a manufacturing company. Instead it is simply a glorified importer, surrendering productive and innovative power to unstable global conditions.

The wealth of the United States comes from planting worthless seeds in the dirt and turning them into life-sustaining bushels of wheat, ears of corn, and high protein yielding soybeans; from disappearing into the side of a West Virginia mountain and coming back up with a lump of energy providing coal; and from taking useless heaps of material and grinding, molding, turning and assembling them into products that make people's lives better. The folks who do this create real, tangible things that add to the overall value of the United States. The rest of the system – the service economy – most often merely shuffles the results of their efforts around, facilitating the wealth creation process or getting goods into the hands of the people who need and want them.

Banking, insurance and legal services are important, to be sure. However, banks just take money from people with a surplus of it and make it available to those with a short term need. Insurance just takes one man's disaster and spreads the cost of it over everyone; and lawyers work to keep the whole system honest – to get the money out of the pocket of the person who has it dishonestly and put it in the pocket of the person who earned it. They keep the social and economic systems well greased and functioning properly, but they do not create much new wealth.

If America ceases to create wealth, there will be nothing left to 'service'. On their worst, most inefficient day, any farmer or production worker makes a greater contribution to the size of the American economy and the economic quality of American life than the most productive lawyer, call center operator or the most prolific Wall Street brokerage firm do on their best day.

Information must also be kept in proper perspective. To the extent that information technology improves manufacturing, it is valuable to all Americans. Information disseminated through books, newspapers, television and the Internet may allow us to grow our knowledge and build our skills. But unless the greater knowledge and skills are put to productive use, the knowledge gained cannot enhance the long term quality of our life; and to the extent that information facilitates faster stock trades, or provides an hour of amusement, it contributes nothing to our economy.

> *When I was young, I heard the older people say, "A shoemaker sticks to his last." Well, my grandfather was a shoemaker and he worked very hard, but died broke.*

The people who actually work in manufacturing – the ones in the factories that see it and touch it every day – know all of this, if not intellectually, then intuitively. They take a feeling of satisfaction and pride in seeing the real physical results of their day's work, eliminating any doubt as to whether their efforts accomplished anything. Manufacturing people rarely spend their retirement years wondering whether the forty years or more they spent on the job meant anything to anyone.

Certainly a stockbroker takes satisfaction from helping an investor make money on a deal, but at some level he also knows that his client's gain is usually someone else's loss and the net of his day's work to society in general was zero. Bankers take pride in helping one get a home loan, but they also know that twenty years from now it won't have mattered to them. However, the guy who builds the house has no such underlying doubt. He knows that twenty years from now the home he built will still be standing, and some family will be safe and warm inside as a result of his work.

Whether or not he knows how it fits into the grand economic scheme of things, the person who actually made something created wealth. Money is simply paper. It only has value when it represents real things – ears of corn or SUVs. When they are produced here, Americans get the value. When they are produced in China, the Chinese get the value. If we have only services and information to offer, and the Chinese, Malaysians and Indians have all of the wealth from manufacturing, they will share that wealth with us only so long as they need to buy our services and information. Once they have their own bankers, stockbrokers, insurance companies and lawyers, there will be no reason at all for them to share their wealth with America. To them we will simply be customers for their products. When the unintelligible language and incomprehensible formulae are boiled off of economic theory, we are left with these simple facts.

Every time an American manufacturing company closes a plant in this country and outsources the work to China, India or Mexico, it makes a difference. The economy suffers and the quality of every American's life goes down just a little. As people transition from manufacturing to service jobs, they typically realize a much lower pay scale with fewer benefits.

> *Outsourcing never made sense to me, but since I was never really in manufacturing, I was afraid to express my opinion - I didn't want to sound stupid. But I could not see the logic in the "virtual company!" I never liked the idea of turning parts of a company over to others, especially the key, value adding parts. Sure, a company cannot do everything. It has to use very specialized service providers, such as the airlines, FedEx, and others*
>
> *But, it just never felt right to me to turn manufacturing over to others. It especially doesn't make sense to outsource HR. I never even liked to let others do the catering! I like to eat well and if I can afford it, why not hire someone from the best cooking school in the country?*
>
> *Toyota sees it the same way. They know they can make automobiles and they knew to make automobiles required bending steel, building engines, and more recently, becoming proficient with electronics. Why should they not be good at all these things? Why not make their own machines while they are at it? Why not learn how to make robots?*

Regardless, the rationalization for outsourcing American manufacturing at its alarming current rate is that it seems to provide the greatest return to the investors. The implication is that there is something fundamentally wrong with America that makes manufacturing elsewhere more profitable – labor costs are too high, environmental and labor laws are too strict, or taxes are too high.

These explanations ignore the 800 pound gorilla sitting in America's living room. More precisely, the gorilla is sitting in Georgetown, Kentucky and its name is Toyota. With American workers and American regulations, Toyota rolls on year after year, gaining market share, racking up profits and providing stable employment.

The big manufacturing companies see themselves caught between the horns of a dilemma, having to choose between what is good for the stockholders versus what is good for America. They are not really caught, however, and Toyota proves that every day. While the stockholders reaping the benefits are Japanese, rather than American, Toyota demonstrates that American manufacturing can be quite profitable. The managers of American companies are not outsourcing because America leaves them no alternative. They are outsourcing because they have been unable to run their companies like Toyota. In the end, they satisfy neither the stockholders nor the country.

For the most part, the great American manufacturers have tried to duplicate Toyota's methods and have launched lean manufacturing strategies. For the most part, they have failed.

Health care and other legacy costs may well be too high for some of the big companies to be competitive. Japan and other countries may well have unfair trade policies. American labor may well be too expensive. There are likely to be all sorts of issues that adversely impact the competitive position of the big manufacturers, but that does not explain why they failed to implement their lean strategies successfully. It is not as though they became lean and achieved quality levels, inventory turnover, and labor hour efficiencies comparable to Toyota, but were still unable to compete due to those factors. It is quite the contrary. They failed to execute their lean strategies. Then the matters of health care, trade policies and labor costs were raised, almost as if to say that it would not have mattered if they had become lean anyway.

> *Years ago, I worked with UCLA Professor Lou Davis who taught a concept called Socio-Technical Systems. United Defense in Aberdeen and the other plants opened by Lou and consultants trained by him operated with self-directed work teams and virtually no layers of management. They typically performed at levels equal to or better than many lean companies. Why we never furthered this concept in America is beyond me!*

Nor can the failure be attributed to anything unique about the Japanese or Asian culture. Toyota runs its vaunted Toyota Production System in Kentucky, as do many of its American suppliers. Honda nearly duplicates it in Ohio. Many smaller American companies have become quite lean. Even within the big publicly held companies there are pockets of lean success. General Motors' NUMMI plant is an excellent example. United Defense' Aberdeen facility (now BAE Systems) is another example of a very lean American company.

Emerson Electric has a few very lean factories. Most of the big companies have at least one lean manufacturing showcase. These plants demonstrate that American workers are fully capable of contributing to a profitable lean manufacturing operation.

Those isolated examples also serve to undermine the notion that there are subtle nuances of lean that American managers have not been able to discern. The country's brightest business students have gone to work for our big manufacturing companies and have read the books written by the Japanese experts and pioneers. It is hard to imagine the problem is an inability to grasp any of the lean principles and techniques. Just in case, however, the big companies have spent millions of dollars in consulting fees to assure professional advice and guidance in the development and execution of their lean strategies. Lack of information and knowledge cannot be at fault.

It is particularly interesting to look at a couple of big companies that have enjoyed at least a partial measure of lean manufacturing success. They seem to be exceptions to the rule. Harley-Davidson is one. The market value of the motorcycle company is getting close to that of General Motors, largely as a result of its Materials As Needed (MAN) lean manufacturing strategy. But Harley-Davidson did not become lean as a publicly held company. It was a poorly performing division of AMF when its management bought it out in 1981. MAN was launched in 1982 and it was not until manufacturing operations were performing at levels equal to or better than their Japanese competitors that the company returned to Wall Street in 1986.

Motorola is another very interesting case. Bob Galvin championed Bill Smith and others who pioneered Six Sigma in the 1980s. It was pure lean manufacturing with a heavy mathematical spin. The guiding principle was the belief that the best quality producer would be the shortest cycle time producer, and the shortest cycle time producer was always the best cost producer. It was wildly successful. Motorola won the Malcolm Baldrige National Quality Award, and the Motorola "Bandit" plant, where pagers were made to order in lead times measured in hours at near perfect quality levels, was the Mecca of American manufacturing. Motorola rode Six Sigma to a big comeback, competing with and often outperforming the Japanese.

Bob Galvin left and, inexplicably, Motorola has all but abandoned Six Sigma as a manufacturing strategy. While the lean companies demonstrate every day the validity of Motorola's quality/ cycle time/ cost principle, Chris Galvin explained that Motorola would be more "agile" and "flexible" by outsourcing manufacturing. At the height of its lean manufacturing success, no other company on earth seemed to be more flexible and agile. Its failure to sustain should lead to some very important questions.

Under Chris Galvin's short-lived (1999-2003) reign, Motorola lost 60,000 jobs and significant market value. Ironically, Motorola University still teaches Six Sigma to others, even though Motorola is rapidly getting out of manufacturing.

Six Sigma cannot be discussed without considering the impact it has had on General Electric. GE is experiencing an enormous level of business success, and it attribute's much of it to its wholehearted commitment to Six Sigma. Like Motorola, however, it is getting out of manufacturing at breakneck speed. GE's various financial services groups are its leading profit generators and GE has a stated goal of outsourcing 70% of all it does, with 70% of that going out of the U.S.

It should be noted that GE has long been a manufacturer, first and foremost. The outsourcing strategy follows two decades or more of failure to succeed in one lean manufacturing initiative after another.

In the first Lincoln-Douglas debate, Lincoln accused his opponent of using "a specious and fantastic arrangement of words, by which a man can prove a horse-chestnut to be a chestnut horse." That seems to be the case with lean manufacturing and Six Sigma. By redefining Six Sigma as an analytical tool to justify outsourcing, rather than a manufacturing improvement tool, a failure to execute a lean manufacturing strategy is redefined as a business success.

Imagine how US workers feel when they are told to learn and install lean when the result brings unemployment for them. I remember years back when I managed a data processing facility and was offered a very large contract from a telecommunications company. When I first looked at the specifications I could not imagine how to do it, let alone expect my operators to be able to. But, I knew we all needed to eat. So, I said, "Look, another company is doing this and if they can do it surely we can too!" Miraculously, we did learn, slowly, but we did learn how to do the work as well as our competitors did – and we all kept eating!

They are not the only folks who can turn horse-chestnuts into chestnut horses. Katrina C. Arabe, a writer for Industrial Trends wrote, "Lean manufacturing is in, while vertical integration is out. That's why companies are outsourcing manufacturing — to stay focused and profitable." These big companies certainly did not tell the world they planned to outsource in order to stay focused and profitable when they launched their lean strategies. They planned to keep manufacturing in their own plants, and to become very lean in order to stay focused and profitable. It has only been since they have all failed in those efforts that lean manufacturing and Six Sigma have been redefined as tools to facilitate the abandonment of manufacturing in America.

And "vertical integration is out"? Lean manufacturing was conceived in the early days at Ford Motor Company, and then taken to an art form by Toyota. Both companies strongly pursued vertical integration. Outsourcing is not an aspect of lean manufacturing. It is quitting.

Clifford F. Ransom II, Vice President of State Street Research in Boston, pays close attention to the subject and he estimates that, at best, one or two percent of the manufacturers with publicly traded stock have actually become lean manufacturers to the degree that the impact can be seen at the bottom line. The rest just talk lean. That miniscule percentage consists of companies like Harley-Davidson or Danaher – a company that was created around a lean vision by its founders after the Japanese manufacturing explosion of the 1970s.

It includes none of the big manufacturing powerhouses that dominated the economies of the U.S. and the world since World War II. Across the board, these companies have announced lean strategies, and then failed to execute their plans.

All logic and reason dictates that at least one, if not more, big American company would have met and perhaps even bettered the Japanese challenge by now. American talent, technology and our manufacturing heritage surely should have enabled one of the great manufacturers to have met the state of the art. Even though there is no other significant field of technical, economic or creative endeavor in which the United States cannot rise to world class levels, manufacturing is quickly becoming one. Of all the areas in which to fail, this collapse of manufacturing capability may well be the one with the most devastating long range consequences. It is no trivial matter that not one of the giant companies – General Motors, General Electric, IBM, Ford, Motorola or the rest have succeeded in becoming a competitive, lean manufacturer.

American managers are not the greedy louts some in the media make them out to be. Nor are they the bumbling ogres Michael Moore would have the world believe them to be. They are very intelligent and capable people, completely dedicated to the companies for which they work and the strategies to which they have committed. It is clear that there is something more to it than the superficial and often disparaging explanations provided by both the companies and their detractors.

If modern managers have a common flaw, it may well be a lack of knowledge about the past. Most manufacturing people actually know very little about Henry Ford, F.W. Taylor, Alfred Sloan or Kiichiro Toyoda and other men who have had an enormous impact on how manufacturing is managed and executed. The tools of the trade, including labor standards, MRP systems, responsibility and flexible budgeting, inventory control and the other standard industrial engineering, materials management, and cost accounting practices are largely taken for granted. We are willing to take on faith that all of those basic manufacturing building blocks are tried, tested, and validated beyond question. We assume that they constitute not only the best way, but the only way, to manage manufacturing.

Several years ago, I arranged for the brilliant industrial engineer behind the Toyota Production System, Dr. Shigeo Shingo, to give a two day lecture at Harvard. I invited a renowned business author and professor from their faculty to attend. He listened for a few hours, and then got up and left. I was amazed. Here was the greatest manufacturing genius of the last sixty years and the Harvard professor did not think it was worth his time to listen. Harvard is regarded as the top business school in the country. The fact that they could not appreciate Dr. Shingo should tell us quite a bit about our business thinking.

Lean manufacturing is viewed as a set of tools, tricks and techniques to be used on the factory floor to supplement these 'proven' and 'necessary' management practices.

Taken together we have an integrated system of managing corporations and factories that has been in place since long before any of us began our careers. We also know that these systems worked well, enabling the architects and masters of modern management to make enormous profits for decades, until the Japanese manufacturing revolution began.

The fact that none of the extremely adept practitioners of these management cornerstones, which have been fine-tuned in the best business schools and pursued to fabulous wealth in the 1950s and 1960s, have been able to adapt to lean competition should give us reason to pause. We should at least be sure we really understand what those management tools and systems are all about, where they came from, and for what purpose they were developed.

In many circles the mere suggestion that Alfred Sloan and his "M Corporation" theory of decentralized responsibility with centralized control might have erred borders on heresy. Dr. Edgar Schein of MIT recently said of Sloan, "His ideas were so clearly correct that we have forgotten that they were an invention." However, it is hard to point to a manufacturing company living by Sloan's principles whose performance over the last thirty years demonstrates that Sloan was so "clearly correct." Sloan's principles are considered to be beyond question, even though his own company – General Motors – has reached the previously unthinkable condition where people are questioning whether it will survive. We have put Sloan on a pedestal. Dr. Schein, who is far from being alone in his high regard for Sloan, is a professor at the Sloan School of Management. The Sloan Management Review is one of the most highly regarded business publications in the world. Even Bill Gates gave Sloan the highest marks in the business class, saying that Sloan's book – *My Years With General Motors* is "the book to read if you are only going to read one book about business."

Then again, Bill Gates has never tried to run a factory by Sloan's rules, and neither have Dr. Schein or the editors and publishers of the *Sloan Management Review*. I have yet to read where a manufacturing manager in his golden years has looked back and credited Sloan or his books with being of much assistance.

While Sloan was brilliant and presided over the growth of the largest industrial enterprise in history, the harsh reality is that his company – General Motors – has been the least able of all of the great manufacturers to step up to the Japanese challenge. It has been the company that most needed to match Toyota improvement for improvement, and it has proven incapable of doing so.

While all of the big manufacturers have adopted the Sloan model, General Motors has been by far the most rigid in adherence to it. General Motors management has always been, and still is, a product of the Sloan system. If Alfred Sloan had been alive and running GM during the 1970s, 1980s and 1990s it seems a very safe bet that he would have done just about as his protégé successors did. It also seems a very safe bet that he would have failed just as dismally as they have.

A brief study of the history of lean manufacturing and the history of modern American manufacturing leads to a couple of inescapable conclusions. The success of the Sloan-managed companies in manufacturing came only during the period from the end of World War II until the mid-1970s when the Japanese came roaring back to life. As a result of the war, those companies had no real competition, except each other. Faced with real competition, no Sloan-managed manufacturing company has succeeded. It is apparent, when the basic principles of Sloan are examined, that the revered Sloan system has never and will never spawn competitive manufacturing. The principles upon which it is based drive companies to pursue bad manufacturing practices.

A Sloan-managed manufacturing company, which includes just about every publicly held company in America, must re-evaluate its entire management structure and systems in order to best the Japanese and the rest of the world.

CHAPTER TWO

The Birth of Lean Manufacturing in the Modern Age

"You've got to go out on a limb sometimes because that's where the fruit is." - **Will Rogers**

Any understanding of modern manufacturing in the United States or anywhere else must begin with Henry Ford. Good and bad, Ford had more impact on manufacturing than any other human being in history. Understanding him personally is not as important as understanding the decisions he made and the consequences of those decisions. In fact, a personal understanding of Henry Ford is, in all likelihood, impossible. Despite fabulous personal wealth and staggering profits, Henry Ford was not motivated by money. This greatest of all capitalists saw himself as a social reformer and detested wealthy people and most of the trappings of wealth. His home, though huge, is stark, notable mostly for the technology built into it. His diet was plain and servants were few. His ideas of a good time were square dancing and bird watching. The power that came with his business success was far more important to him than the creature comforts he could afford.

Henry Ford was not a "car guy." The man who led the design and manufacture of the most influential, profitable and widely sold car in history would have taken no particular joy from tooling down a winding mountain road in a modern, high performance car. The early auto makers solicited backers and customers by racing their inventions against each other. Ford drove his own car in a race once, and won. From then on he was content to hire someone else to race the cars he built.

Nor would he have taken enjoyment from the luxury and comfort of modern climate controlled, surround-sound filled, ergonomically designed vehicles. Cars were utilitarian devices to Ford, much like tractors or steam irons. They were simply the tools it took to accomplish a purpose as expeditiously as possible.

For a social reformer, Ford was not reported to be a particularly amiable man. He believed hard work was its own reward and that all a man needed in life was a worthwhile job at a fair rate of pay and a home in which to spend time with his family.

Perhaps the most absurd of all myths concerning Henry Ford is that there was no organizational structure, and that he trusted no one, personally directing all of the activities of the company. This amazing group photograph includes over 12,000 employees working at Highland Park shortly before the advent of the assembly line. As with many people in large companies, most of them never personally talked to Ford over the course of their careers.

For the most part, Ford said what he meant and meant what he said. The often misrepresented 'five dollar day' was nothing more than a profit sharing plan. He believed that he and his managers and the other stockholders had taken more than enough of the profits. He had given profits back to the customers in the form of continuing price reductions. It simply struck him as the most reasonable thing in the world to give the rest of the profits back to the employees.

Ford was no deep thinking, economic genius. He was likely the most surprised of all to find that reducing prices and increasing wages generated even greater profits, even though he later claimed that he had planned it that way all along.

> *My father hated Henry Ford for his attitude toward Jews. I carried his feeling with me until one day in Japan I asked Taichi Ohno, VP of manufacturing at Toyota, how he discovered JIT. He laughed and said, "I simply read Henry Ford's 1926 book Today and Tomorrow." When I returned to Connecticut, I asked a Greenwich librarian to find a copy of the book for me, which she was able to do. I read it and was so fascinated that I asked the publisher, Doubleday, for reprint rights. By the time I left Productivity Press in 1999, we had sold 35,000 copies of this book published in 1926, making over a half million dollars. I felt my father would have had a good laugh and forgiven me for publishing and profiting from Henry Ford's book.*

The success of the Model T convinced Ford for a time that his values were shared by most of the world, and that he was indeed helping the rest of it move closer to his Calvinistic outlook. It also seemed to have convinced him that he was wiser than turned out to be the case. Pollyannish ventures such as the peace ship were launched, as well as his vile, ridiculous ranting against the "International Jew."

In the end, reality set in. It came hard to Henry Ford to realize that people wanted more from their cars than a means to get from point A to point B; they actually wanted to enjoy them. He felt betrayed by employees who had higher expectations from life than a chance to work hard and sit at home. He was disappointed in himself and in his son, when it turned out that Edsel enjoyed the nicer things in life and the company of perhaps frivolous, but interesting people. Ford became bitter and at times vindictive. In his final years, he was not known to be a particularly good manager or a good man.

Generally, Ford stayed out of the factories after the Model T reached its peak. When the company moved to the big plant at River Rouge, he kept his office in a separate engineering building. Charles Sorenson largely ran things from the Rouge and did a pretty good job, for the most part, of keeping Henry Ford and his trouble-making cronies at arms length from the day-to-day business of manufacturing.

It is important to try to keep Ford in perspective. While he was personally erratic, the period from around 1908 when the Model T was first developed, to about 1918 or so when Ford became more Don Quixote than manufacturer, were extraordinary years. Even though Henry Ford was at odds with the values, background and character of a 21st century manufacturing manager in just about every respect, there are still valuable lessons to be learned. It is important to focus on actions and results; the management decisions that gave birth to lean manufacturing and not allow them to be diminished by either incongruent motives or later year eccentricities.

There is a popular misconception that the origin of the assembly line was the product of a sort of "Eureka" moment that Henry Ford experienced, and that it was inspired by an observation of meat hooks passing by on a moving overhead chain in a Chicago slaughterhouse. Nothing could be farther from the truth. Mass production and the most visible assembly line – that at final automobile assembly where all of the disparate parts and subassemblies come together to actually become a car – were the products of a strategy that drove hundreds of simultaneous projects, only one of which was that final assembly line. If there was any direct inspiration, it might have been the crude assembly line used by Ford's closest competitor, Ransom Olds, who was cranking out his little 'curved dash' Olds in crude assembly line fashion before Ford made the Model T.

The early Ford Motor Company was an absolute hothouse of ideas and change in the manufacturing processes, especially during the period from 1910 or so to about 1915. Charles Sorenson, the Assistant Superintendent of Production at the time, said quite accurately that Henry Ford was not the father of mass production, but the sponsor of it. By that he meant that Ford had little personal involvement in most of the continuous motion projects going on at the time. What Ford did was create the environment that drove dozens of people to dream up and implement those projects. This is an even greater compliment to Ford than assigning him credit for engineering a few moving belts and chains. Sorenson's assertion gives Ford credit for management and leadership.

Henry Ford's personal claim to process engineering fame is limited to a moving conveyor upon which magnetos were assembled. Even there, most of the grunt work and analysis were done by a foreman named James Purdy, who was in charge of that corner of the plant. Ford took an interest in the project primarily because the Model T magneto was one of a few components in the car that he had completely engi-

Charles Sorenson was the draftsman for the Model T and Assistant Production Manager at Ford when the assembly line and lean manufacturing were born. In 1929, he taught Ford's manufacturing methods to Eiji Toyoda. Early in World War II, Sorenson visited the Army's bomber plant. He saw the process, worked all night in his hotel, then presented the Army with detailed plans for a production process that soon became Ford's Willow Run plant, turning out 8,685 B-24 Liberator bombers. He was with Ford longest of the early managers and was fired in 1944. He went on to run Willys Overland. While Ford's behavior became more erratic, Sorenson stayed above it. He ran the River Rouge plant with a firm hand, avoiding the political storm that swirled around Ford. Harry Bennet, Henry Ford's strong armed assistant who did much of Ford's more unsavory bidding, left Sorenson alone. He recounted that Sorenson once told him that he could "whip six Harry Bennets." Bennet wasn't sure about six, but he was certain that Sorenson had no fear of one Harry Bennet

neered by himself. The bulk of the motion projects were driven by foremen and engineers who are lost to history.

Just about every change in the flow came from the bottom of the organization up, rather than from Henry Ford down. If credit must be assigned to one person, that person would be Clarence Avery. Avery held the job title of Foreman, but was a foreman without production responsibility. As a sort of special projects/staff person working for Sorenson, he, more than anyone pushed, urged, coordinated and otherwise shepherded all of the flow projects into some semblance of unison.

A common saying among the workers in the shop at the time was, "Everyone takes the hardest way first." There can be no clearer evidence of a continuous improvement culture. The workers all understood that, however work was being done, there was an easier, better, and faster way that could and would evolve.

> *Once while visiting Toyota, I turned to Ohno and said, "I am very grateful that you allow my study missions to visit Toyota, but why is it that every plant we visit looks old?" "Bodek-san, you do not understand the Toyota Production System." It was a classic answer he often gave me. I was at first confused with his statement until I realized that Toyota does not spend money 'foolishly' on new machines. Even today, they make 90% of their own machines. I subsequently learned that a new machine, in Toyota's opinion, was in the worst possible shape. Looking new had nothing to do with it. The question is, "does it work and does it work well"?*

The first system that truly integrated material handling with the production process was in the foundry in 1912, the result of a suggestion by a mechanic named Gregory whose inspiration was a process he had seen in a brewery. The motor assembly line was suggested and driven by William Klann, the supervisor of that area. It seems to have been almost contagious. Projects resulting in conveyors, chains, and chutes sprung up throughout the plant.

The final assembly line conveyor was almost an afterthought – but one in which plant management had little choice. The motion projects in all of the primary manufacturing and subassembly areas raised throughput so radically that final assembly was the clear bottleneck. The plant was capable of flooding final assembly with four, five, six times as many parts as it could handle.

While seeing cars come together is fascinating, and the press and the public stood in awe of the intricate ballet of perpetual motion that seemed to make assembled cars arise from nothing every few minutes, true manufacturing professionals focused elsewhere. Ford and his management all understood that mass production required three elements in this order: quality, cycle time (they called this "motion") and synchronization.

First, was quality. The term used by Ford and his manufacturing staff was more often "accuracy." Manufactured parts had to be 100% in conformance to specification. Any variation in quality not only meant that the defective part would not work, but that the entire flow would come to a halt.

Second was cycle time or "motion." All parts and subassemblies had to be on the move, headed in the right direction all of the time, slowing down or stopping from time to time just long enough for work to be performed. Third, they needed synchronization, or "precision" in 1910 Ford lingo. These were the three pillars: Quality enables flow and flow requires synchronization. The first of these, quality, was the real engineering breakthrough and it was Ford quality that awed the manufacturing community.

It is important to make the distinction between manufacturing quality and quality to the customer. Model T customers were, as a rule, very pleased with the car. However, the Ford files have plenty of letters from early Model T buyers complaining or pointing out problems. The distinction is that the letters almost universally complain about design issues – the cooling system was inadequate, a spring that had been specified to be made from a steel that was too brittle, and so forth. It was very rare that there were complaints about how the manufacturing process performed.

This is apt to be the inspiration for Toyota's "bottoms up" management style. It is based simply on the concept that the worker doing the job knows that job the best, and if you want to make improvements it makes sense to first ask the "expert," "What do you think?" Involving the worker in problem solving activities is critical to the success of Japanese management.

When the assembly line and production of the Model T really cranked up in 1914, there were over 15,000 machines in the plant, many of them designed by Ford mechanics and engineers. Each engine, chassis and drive train component required up to a dozen or more machining operations to very close tolerances. The real manufacturing wonder at Ford was the performance of these thousands of grinding, boring, milling and lathe operations every day with consistency and precision, and with a very high degree of machine reliability. There was no room in continuous flow for machines to break down either.

The production managers controlled preventive maintenance, while another fellow lost to history – Carl Emde – ran a department called Tool Design. Emde took over after Oscar Bornholdt left in 1914, but Bornholdt left behind a track record and an example of machine design that could all but assure accurate, high quality production. Today their group would be called Manufacturing Engineering. First Bornholdt and then Emde, and some 59 other men made Ford machining quality happen. Success in achieving accurate machining with continuous flow on the scale of the Highland Park plant was the real manufacturing marvel of Ford. Assembly lines were the logical next step and a comparative walk in the park after that.

While there is no question that Henry Ford completely missed the signs that the market was changing, there has never been a manufacturing operation in America comparable in terms of speed, quality, or cost on the scale of Ford in its heyday.

The company was so intent on manufacturing excellence and so driven by Ford's misguided belief that cost and utility were the only aspects of cars that mattered to consumers, it nearly went under. To the end, however, there was no questioning the manufacturing results. Although fewer and fewer people wanted them, the cars that came out of Henry Ford's plants were high quality and had been made at a very low manufacturing cost.

Business writers and many academics - particularly those whose purpose is to sing the praises of Alfred Sloan - make much of the Ford Motor Company's accounting and administration in the 1930s when Henry Ford was both the sole owner of the company and at his personal worst. They should be looking at Ford during that period when lean manufacturing came to life, and at the management systems and practices in place then.

James Couzens and Henry Ford were equally responsible for the birth of lean manufacturing and the Ford Motor Company. While Henry Ford was responsible for engineering and production, Couzens ran the Business Department, including sales, accounting, purchasing, personnel and administration. Couzens and Ford's relationship was rooted in mutual respect than friendship.

Ford's partner, James Couzens, was responsible for the Business Department, while Henry Ford ran manufacturing and engineering. There was nothing undisciplined or shoddy about management under Couzens. While he may have wreaked havoc on accounting from time to time in the later years, Henry Ford did not dare to tamper with accounting while Couzens was in charge.

Couzens was responsible for accounting, sales, purchasing and human resources. Each function was well designed, responsibilities were well defined and Couzens ran his area with a firm hand. Most important, Couzens created and controlled an accounting and cash management system that demanded lean manufacturing and enabled staggering growth.

While Ford was driven by production flow, Couzens was even more driven by cash flow. Typical supplier terms called for payment averaging 25 days after the receipt of material at the plant. The manufacturing cycle was a week or less, shipping time another week or so. Payment terms to dealers were C.O.D. The result was that Couzens had the money from the sale of a car in the bank a week or more before the payments for the parts that went into it were due.

CHAPTER THREE

The Industrial Engineers

"The human being always looks down when he is examining another person's standard; he never finds one that he has to examine by looking up."
- *What Is Man?*- **Mark Twain**

Most Americans, if they know of Frank Gilbreth at all, know him as the quirky father in *Cheaper By The Dozen*. Some may even remember him as an 'efficiency expert' constantly looking for the most efficient way to do everything from bricklaying to buttoning shirts. Both views of him are true, but Shigeo Shingo, the Japanese industrial engineering visionary saw him as much more than that. It is no coincidence that Shingo cited Frank Gilbreth in his books, but makes no mention of F. W. Taylor, who is regarded as the greatest of all the 'efficiency experts'.

Taylor was the first, and the originator of scientific management. In 1903, the same year the Ford Motor Company was incorporated in Detroit, Taylor published *Shop Management*, the first all-encompassing description of his theory of scientific management. Gilbreth, twelve years Taylor's junior, in some regards followed in his footsteps. When they first met in 1907, Gilbreth saw it as a great moment and was convinced that he had met a great man. While he always maintained a fair degree of respect for Taylor, it was not too long before the pedestal upon which Gilbreth placed him began to wobble.

FrederickTaylor, the father of scientific management, was brilliant and devoted to his cause, which he sincerely believed was in the working man's best interests. Respected for his work, he had few genuine friendships. People found him abrasive and egotistical. Working people certainly did not appreciate his efforts 'on their behal'. Biographers have noted that none of the working men Taylor proclaimed to empathize with attended his funeral.

Taylor was a stopwatch man, while Gilbreth was a methods man. Certainly there was a great deal of overlap and they were both guided by a star they each felt led them to "the one best way" of doing things. Gilbreth often made films of people at work with an oversized stopwatch sort of device in the background so he could see the time elements of various work motions. Taylor's concern for methods was erratic. At one steel company he had ten or more different shovels designed based on his assertion that a man shoveled best when he was shoveling 21 pounds, ergo a different shovel for each material so that there would be about 21 pounds per scoop. In other cases he left the methods up to the worker and was content to just time them. Both Taylor and Gilbreth had enormous egos, as well, and extended the 'one best way' principle to themselves – they each thought they knew the 'one best way' to apply scientific management. Compromise was not among either man's strong points.

Frederick Winslow Taylor came up

through the ranks as an apprentice and ultimately became the supervisor of a big machine shop. (Some contend that, perhaps through his family connections, he rose a little quicker than his performance justified, but he did work at the bottom of the mechanics' totem pole for a few years.) His first foray into timing work and creating the principles of scientific management evolved as he struggled to take control of an inefficient machining operation. His concept of scientific management covered a lot of ground – far beyond stopwatches – but stopwatches were really the centerpiece. He also firmly believed in separating the thinking aspect of work from the doing. Management should do all of the thinking and planning, while workers performed the doing. There was no need for workers to think at all in a well managed shop, according to Taylor. Instead, Taylor advocated creating positions for nine "functional foremen" who could do all of the thinking, planning, and controlling necessary to get the most out of a good-sized work crew.

Workers ran the gamut from "First Rate Men" at the top of the labor hierarchy, to "Stupid Men" at the bottom. Exactly what made one man first rate and another stupid was a little vague. Taylor apparently assumed it was a matter of common knowledge. (He did give clues, however. Pig iron should be handled by a man ". . . so stupid and so phlegmatic that he more nearly resembles in his mental make-up the ox . . ." and ". . . so stupid that the word 'percentage' has no meaning to him, and he must consequently be trained by a man more intelligent than himself." Work such as "dirt handling" should typically be assigned to "Hungarians and Italians," according to Taylor. While it is certainly offensive, not too much should be made of this outrageous point of view. In all probability, many Italians and Hungarians probably held folks of Taylor's lineage in similar contempt. As with Henry Ford and the nonsensical articles about the "International Jew" in Ford's newspaper, *The Dearborn Independent*, these were bigoted men living in bigoted times and such views were common. It is no excuse, but neither man had people around him expressing revulsion at such attitudes. In fact, many of

Frank and Lilian Gilbreth. Frank Gilbreth once said that no man can truly understand management until he has faced making a payroll every week for a couple of years.

After his death, Lilian Gilbreth went on to become Purdue's first female engineering instructor.

their peers shared similar opinions. A "First Rate Man" seems to be any man who would put up with harassment by Taylor and his stopwatch while he was trying to work.

At any rate, Taylor focused quite a bit on eliminating the distractions that kept people from working – going after material or tools, having to scrounge around for drawings and specifications, and so forth – but he largely left it to the men to determine the actual work method. He was interested in seeing how much time it took a "First Rate Man" to do a job, then setting that as a standard. Management's role was to knock down all of the obstacles to production, such as making sure the materials, tools, and drawings were ready. Then it was the worker's job to get the work done in the allotted standard time.

Taylor was clearly focused on individual operations, rather than on overall flow. He sought to eliminate waste, defined as anything that kept a direct laborer from producing 100 percent of the time. That he gave scant attention to process was really a result of his background. He conceived of Scientific Management while working at Midvale Steel and Armament in Philadelphia, primarily a maker of rail for the railroads, striving to get into the cannon barrel business. Midvale was essentially a steel mill with one or two step basic machining processes. Taylor's major consulting clients were Bethlehem Steel and the Johnson Company, also steel and machining companies. In between he built, ran and failed to make money with a pulp mill. None of his work required assembly, flow or synchronization. For the most part, he knew manufacturing as a batch process with perhaps one significant value adding step, then shipment. There is no evidence that flow or synchronization ever crossed his mind.

Gilbreth came at scientific management from another angle. He started life out as a bricklayer and was struck by the fact that, during his apprenticeship, each master bricklayer gave him different instructions as to the best way to lay brick; and none of them actually followed the methods they proscribed for him. He eventually went on to own a construction company and continued to pursue the idea that, out of all of the so called "best ways", there must be one best way to lay brick. He eventually invented special scaffolding and implemented work procedures that reflected the conclusions he reached concerning the best methods. While he enjoyed a fair amount of success as a building contractor, he eventually moved on to consulting engineering, convinced that his business success was the product of his 'one best way' thinking.

Unlike Taylor, Gilbreth concentrated more on methods than time. He began analyzing work in infinite detail, often taking movies of people working and then breaking the work down on film into its least bits for microanalysis. His forte was minimizing the number of motions a worker made, reducing work to its simplest form. The resulting time was not particularly important to him. If a man was working in the most efficient manner possible, then however long it took was how long it took.

That focus on time as opposed to methods was the crux of the personal clash that developed between the two men. Taylor thought that methods without scientifically calculated times were useless. Gilbreth thought that the time to do something was a meaningless figure if the methods were not the best. They clashed in a very civil manner before the various engineering societies, but Taylor was the father of the scientific management movement and the nationally recognized guru. Gilbreth was something of an upstart, so it was a rather one-sided fight.

In 1924, Gilbreth died quite unexpectedly at the age of 55. He had only outlived Taylor by nine years, so there was actually little time for him to operate out of the great man's shadow. His wife and business partner, Lillian, carried on his work after his death, but she was severely restricted by her gender. The country was still a long way away from generally accepting female consulting engineers. Her real interest was in the area that would eventually become known as ergonomics, and she soon left consulting to take a teaching job at Purdue. Gilbreth's work was largely left untended, but not unnoticed.

More reasonable people than either Taylor or Gilbreth quickly realized that time and motion were not exclusive of each other and that, while there was probably no tent big enough to hold Taylor and Gilbreth's egos, the industrial engineering tent was plenty big enough to hold both men's principles. In 1930, Dr. Harold "Mike" Maynard quit his job at Westinghouse, where he had studied and applied the principles of both men, and formed his own engineering company, and eventually the Maynard Research Council. Maynard knew that he could determine the best methods, a la Gilbreth, put a stopwatch to the result, a la Taylor, and he would have the best of both worlds. The combined approach was called Methods Time Measurement, or MTM, and the core principle of industrial engineering for the 20th century was in place.

But there was a more fundamental difference between the two men than time versus motion that was mostly unnoticed in this country, or at least deemed to be insignificant. That difference was process focus versus operation focus. Taylor looked at direct labor and machines, timing them meticulously and setting standards accordingly. Gilbreth looked at the same things, but he also looked for wasted motion in between. Whether it is true or not, Gilbreth laid claim to being the inventor of the flow chart. In the days before computers with 3-D modeling capabilities, he built boxes that represented scale models of factories to show management how to rearrange machines in order to eliminate steps along the entire material flow. His background in construction as well as a number of assembly processes, including a brief stint at Packard, certainly played a part in giving him a broader manufacturing perspective.

While the American industrial engineering community zeroed in on Gilbreth's ideas about motion, his work on flow was for the most part treated as a side issue. Flow charting and Gilbreth's symbols for charting various process elements were adopted by industrial engineering, but treated as good project tools, rather than core techniques for optimizing manufacturing. Shigeo Shingo, who would emerge as Japan's greatest industrial engineer and would eclipse both Taylor and Gilbreth in reputation, did not seem to be as concerned about motion, but he was very interested in what Gilbreth did in the area of flow.

That Toyota would look to Gilbreth and his flow ideas, and that Gilbreth would have a flow perspective in the first place, are not really surprising when the backgrounds of the two men are considered. Taylor was a hired gun all his life, with the exception of a foray into the papermaking business that failed. He was actually more of a contract manager than a consultant, as we would view him today. He had relatively few significant engineering consulting jobs, but they tended to last for a few years or more. He believed it took anywhere from two or three years up to five or six years to implement scientific management fully, so his jobs were planned to last that long. He also had terms that granted him almost full authority to run the plants in which he worked. He was, in many respects, the precursor of the professional manager. He came from a wealthy family, but became well off on his own hook as a result of a patent for high speed cutting steel he co-owned with Bethlehem Steel. Cash flow was never much of a consideration for Taylor personally or professionally.

Gilbreth, on the other hand, had owned his own construction business, which he started with virtually nothing. He developed a reputation as a "speed builder." That is to say that he could put up a quality building much faster than his competition. Speed was essential to him, not because he was particularly concerned about speed itself as he was about getting paid. Those were days before progress payments were typical. His work simplification efforts kept his cost down but, more important, they shortened the time until he was paid. According to Frank Gilbreth, "No man can understand management until he has looked a payroll in the eye every week for at least a few years." Gilbreth concocted an elaborate set of charts and odd symbols, along with his scale models, that were actually the first process maps. They were designed to help management determine the shortest path to payday. In Gilbreth, rather than Taylor, Shigeo Shingo found a more kindred industrial engineering spirit.

Shingo gave Taylor his due. "In 1931, I ran across a translation of Taylor's book *Principles of Scientific Management* in a neighborhood bookstore. Thumbing through it, I found a most unusual statement. 'Inexpensive goods,' it said, 'can be produced even when workers are paid high wages.' The apparent impossibility of such a proposition aroused my suspicions, and as I continued to leaf through the book, I saw that Taylor claimed the feat was possible if efficiency was raised to a high level. For me, this argument was utterly novel, so I bought the book and

did not sleep until I had read it from cover to cover. At that point I resolved to devote my life to scientific management."

Fortunately for Toyota, Shingo did not stop learning when he read Taylor's book. He studied all of the scientific management experts, Gilbreth included, as well as the books written by a couple of Japanese men with unique points of view. Most likely, Shingo read Ikeda Toshiro's *Secrets To Omit Useless Procedures (also translated as Secrets To Omit Unprofitable Ideas)* before he read Taylor. Tohiro's book is a sort of "Horatio Al-

Shigeo Shingo, the great Japanese innovator redefined industrial engineering as a strategic activity, rather than a technical discipline. He had a keen sense of humor and a dry wit.

When he sustained a serious head injury, a doctor had to drill a hole in his skull to relieve the pressure. Long after he had fully recovered, the hole remained. He quipped that his insights resulted from having a brain that was "better ventilated" than most people.

He said that Japanese people were very dumb, and that Americans were very smart. But often they are so preoccupied with great thoughts that they make dumb mistakes. In both cases, work had to be kept very simple.

Finally, he said that Japanese keep things simple, while Americans make things complicated. That explains why Japanese have thick, black hair, while Americans have thinning, gray hair.

gers meets Frederick Taylor" work of fiction. The story line is that a bright, hard working young boy discovers the principles of scientific management and uses them to better the lives of all Japanese people and, in the course of enriching the lives of everyone, becomes a national hero. It is hard to imagine how any young Japanese boy would want to become anything other than an industrial engineer with inspiration such as that!

Shingo also read the ideas of Ueno Yoichi, who was something of a scientific management maverick. His concepts blended a touch of philosophy with Taylor, and he urged the need to find a time standards balance point. If the standard was ten minutes, for instance, Yoichi preached that it was unfair to management for the worker to take longer; but if management needed only one piece it was unfair to the worker if management wasted his time having him make two. Both worker and manager needed to be careful not to steal time from each other.

Shingo had a brief pre-War stint on the industrial engineering staff of the Taipei Railway, implementing scientific management principles, but most likely his greatest educational experience came from the school of hard knocks. During World War II he left the staff and theoretical realm and worked as a production manager in an armaments plant. There he had ample opportunity to see all of the principles and theories in action, and to devise for himself his own perspective on the subject.

A year after the War, Shingo presented the Japan Management Association with a paper entitled "Production Mechanism of Process and Operations", in which the relationship between operations efficiency and process performance were explained and the rest, as the saying goes, is history.

While Shingo was clearly inspired by Taylor with the potential of industrial engineering, there was not much evidence of Taylorism in his results. Shingo, by the time he was finished, became the greatest industrial engineer of them all. His work in quality control and absurdly short machine change over and set up times enhanced manufacturing far more than Taylor's work.

Pure Taylorism in Japan was somewhat a casualty of war. In the early days of scientific management in Japan, Gudo Takuo was an influential force and the industrial engineer with perhaps the most exposure to American applications. He advocated time standards to be sure, but also a separate and powerful staff functioned to do the thinking in the factory. He found his most willing audience in the growing Japanese military during the build up in the 1920's and 1930's that led to World War II. That Taylor-centered view disappeared from Japan along with the rest of its military structure when the War ended.

Most important, as his 1946 Japan Management Association paper made clear, Shingo realized that attention to both process *and* operation was essential. He realized that work should not be performed at the *fastest pace possible*, but the *exact pace necessary*. Shingo taught thousands of industrial engineers the critical nature of takt.

Takt is the pace of demand. For example, if the plant runs fifty hours per week and the demand is for 500 units per week, then the plant must produce ten units per hour, or one unit every six minutes, to keep up with demand. Industrial engineers, Shingo saw, should devise processes that enable production at one every six minutes, using the least labor possible. This is fundamentally different from the Taylor approach. Taylor would have each operator produce 500 as fast as he or she humanly could, then sort out the pieces and put things together later.

In effect, Shingo created an invisible assembly line. The entire plant worked at the pace of customer demand whether there was a physical conveyor regulating them or not. While the American industrial engineering community was forming in lockstep behind Maynard's blend of Taylor and Gilbreth, focusing its efforts on isolated operations within factories, Shingo had seen the factory as a whole and devised a vision of industrial engineering that defined process optimization as the overarching objective. This was perfectly in keeping with the vision at Ford when the assembly lines sprung to life in 1914. The measure of operations was how well they enabled process optimization, and the operations improvement tools and techniques supported that goal. This global view of manufacturing far surpassed anything being done by Shingo's American counterparts other than, perhaps, the early Ford team.

There was another marked difference between Taylor and Gilbreth that has passed unnoticed, but really defines the differences in managerial approach that were evolving in the early days of modern American manufacturing. Taylor viewed workers very much as liabilities. Left to their own devices, through ignorance or laziness, they would not get much done in the course of a work day. Accordingly, they required clear direction and little, if any, room for influencing the content or pace of their work. His approach was essentially to put the workforce in the smallest box possible, and then assure the maximum output from them. He proposed the need for nine "functional foremen" to see to it that the work was being performed properly, including a "gang boss", a "speed boss" and a "disciplinarian." The notion that a manufacturing business has need for a full time "disciplinarian" speaks volumes about Taylor's view of workers.

Needless to say, F.W. Taylor was not too popular with the working people of America. Labor unions, in particular, viewed him as the enemy. For a long time, the use of stopwatches in any plant performing work for the government was prohibited in direct response to Taylor and scientific management. Strikes and walkouts were common when manufacturers Taylorized. Ultimately, however, a government sympathetic to the growing manufacturing community cleared the way for Taylorism everywhere.

> *The great learning for me was to see Dr. Shingo stop and look at the process, take out his stop watch, and then ask simple questions of the engineers. Imagine the progress you could make if this teaching process were duplicated daily on the factory floor?*

Gilbreth and his wife had a much more positive way of looking at things. They believed that the average working person very much wanted to provide a day's work for his day's pay, but was prevented from doing so by poor design of the workplace and the work itself. In one of his books, Gilbreth asserted that if the foreman knew what every worker was doing all the time, he must be "neglecting matters of much more importance."

The Gilbreth's approach was to make work easier to perform. The Gilbreth criticism of Ford's assembly lines, for instance, was that the height of the line was based on the height at which the average man found it physically easiest to work. By definition, therefore, it was a little too high for some and a little too low for others. They took no issue with the concept or the speed – just the fact that any fixed conveyor or assembly line put more back strain on workers than individually designed work stations.

The Taylor view of people was basically negative, while the Gilbreth view was more positive – or at least accepting of the idea that people were a necessary input. Taylor saw people as a necessary evil. Gilbreth saw them as a primary source of manufacturing results. Taylor wanted to control people, Gilbreth wanted to enable them. Gilbreth experienced only one instance of a labor union walkout, and that was short lived. After Frank Gilbreth died, his wife openly refused to work for any company that she felt was trying to use her husband's methods merely to increase efficiency, rather than to improve the overall flow and factory environment. Often as not, her consulting contracts put her in the role of arbitrator between management and the workers as a result of the respect the Gilbreths enjoyed from both sides.

At first, Dr. Shingo called his system Bakta-Yoke, (fool-proofing). after the name often used in America. But, one day in the factory a woman actually cried when Dr. Shingo spoke about Bakta-Yoke devices. "Why are you crying he asked?" "Because I am not a fool," she replied. At that moment Dr. Shingo changed the name to Poka-Yoke, (mistake—proofing). Poka-Yokes are simple devices created by workers to absolutely prevent defects from occurring. United Electric in Massachusetts built hundreds of such devices.

Taylor's liability view dovetailed perfectly with the system of management soon to take form at General Motors. They were both based on the idea that management should do all of the thinking, while factory labor should do the work. One of the most essential functions of management, Taylor and Alfred Sloan would have agreed, was the need to control the workers. Standard times and standard costs were two peas in a control pod. The system took from Gilbreth only the notion of elimination of unnecessary motion; and that was to save time, not back strain.

Where Shigeo Shingo came out on the matter of people is a little more difficult to discern. He managed largely to avoid the turmoil in Japan in the immediate post-war period when the labor movement was at its peak. He seems to have had a rather matter of fact point of view.

Taiichi Ohno, with a touch of sarcasm, thought people only came in whole numbers; he did not understand the American accounting concept of .1 people. While Dr. Shingo was also a difficult "task master", demanding that people "Do It!", he also had a more humanistic approach to people.

Dr. Shingo was the master at Toyota of Gilbreth's work. I recall walking with him in a factory with a group of engineers. We came to a punch press and watched a worker bend to his left, pick up a sheet of metal, lift it up, shifting to the right, place the sheet into the bed of the press, move his hands outside of the press to touch both safety buttons to allow the press to come down and form the metal. After the metal was stamped the worker reached into the press and removed the formed metal, placing it at his right side on the floor. Dr. Shingo took out his stop watch and asked the group of engineers to determine the value adding to non-value adding ratio. We all watched the process once again.

Dr. Shingo said, "What is the percentage of value adding to non-value adding?" One engineer said, "100 percent, the worker is continually working." Another engineer said, "50 percent." A third engineer said, "30 percent." Dr. Shingo looked at his watch, laughed and said, "Only 14 percent of the time is value adding. Only when the press is forming against the metal is it value adding. The rest of the time is considered as waste or non-value adding."

He looked again at the engineers and said, "What can you do to improve the value adding ratio?" One engineer quickly said, "We could install a leveler to automatically raise the sheets of metal as one is moved into the press." Another engineer said, "We could put the metal on a large table nearer the press and the worker could simply slide the metal into the press." The third engineer said, "We could put a spring at the back of the press that would force the sheet to move forward after the metal is stamped speeding up the removal process."

Shingo was pragmatic in just about every area. He was quick to explain in great detail the practical things he had seen and accomplished and his ideas concerning factory floors, production people and machines. While he was clearly steeped in the theory of the Toyota Production System, he had little interest in going on about theory at great length. He would much rather talk about machines than strategy. To some extent he was limited by his status as a consultant, rather than employee. He was bound by certain confidentiality requirements. He did say, in typical Shingo fashion, that if you want to master Toyota-level manufacturing you "have to grasp not only Know-How but also Know-Why." He was making the point that a large number of American companies seeking to become lean manufacturers had mastered many of the techniques – the

It was a privilege for me to know and work with Dr. Shingo. His writing and work on process improvement right after World War II were the forerunners of the Toyota Production System - lean manufacturing. Dr. Shingo recognized then that manufacturing could be divided into two planes, or axes, one of machines the other of process. Managers and engineers would walk the plant and see series of machines producing products and try to maximize the production of each machine. Dr. Shingo saw the machines on one axis and the total process on another axis. Of course, it is important to maximize production of each machine but it is even more important to focus on the overall process to insure that the customer receives their products on time. Focus only on machines creates great waste as we overproduce.

I once asked two of Ohno's associates, those who brought the Kaizen Blitz to America, "Who created Just-in-Time, Dr. Shingo or Mr. Ohno?" Their answer was, "Who came first the chicken or the egg?" Years later, I asked Dr. Shingo the same question. His answer: "I did for I was Ohno's teacher."

Know-How – but still had no idea why they were doing it.

For the most part, the early industrial engineering and scientific management body of knowledge came to life without influencing Ford, or Ford influencing it. While the Gilbreths toured one of the remote Ford assembly plants, none of the main players in the movement made a study of Highland Park, and Ford management never saw reason to ask Taylor, Gilbreth or their followers for help. Given the magnitude and nature of the manufacturing process breakthroughs being made at Ford, such an attitude is understandable. There was not much Ford could have learned from Fred Taylor or Frank Gilbreth that would have been of much value.

Ford most certainly would have agreed with Taichi Ohno's assessment of industrial engineering. According to Ohno, the real mission of industrial engineers is the improvement of the overall system of production. He classified time and motion studies as merely "the work of technicians."

CHAPTER FOUR

Toyota Takes Charge

"When you do the common things in life in an uncommon way, you will command the attention of the world."- **George Washington Carver**

There was never anything like Ford manufacturing in the world – until Toyota in the 1950s. The Toyoda family adopted the Ford system lock, stock and barrel, but they faced to two significant problems with it. The first was how Ford approached that third essential element of mass production – synchronization.

The most thankless job in the world must have been that of the man called the Shortage Chaser in Ford's Highland Park plant. The typical cycle time from raw material in the back door to a finished car rolling out to the railroad siding for shipment was about four days. One hundred freight cars of material came in every day, and a thousand or more automobiles went out during that same period. Despite everyone's best efforts, there were production problems as all of the hundreds of manufacturing processes flowed together to ultimately meet at final assembly. The Shortage Chaser was the man charged with keeping on top of it all. He held forth in an office on the shop floor called the Clearinghouse, and he had a small gang of Checkers and Counters working for him. The

Checkers and Counters constantly (as their job titles suggest) checked production volumes and counted parts and subassemblies throughout the plant and reported back to him. The Shortage Chaser kept tally on blackboards in the Clearinghouse and alerted Production Superintendent Pete Martin or his Assistant Charles Sorenson if any shortages were imminent. In these pre-computer days, this brute force clerical approach to production control and synchronization seemed to be the only way that worked.

Taiichi Ohno had a reputation for being fair and brilliant, but demanding, tough, and impatient. It is clear that he was also stubborn.
The Just In Time flow concept, built around kanbans, took twelve years of teaching, cajoling, demanding, and a great deal of trial and error.
He began as a machinist in the Toyoda Loom Company and joined the car company in its infancy, first as a machining manager. He eventually rose to Director of Manufacturing.
Towards the end of his career, Toyota sent him to Toyota Gosei, a supplier company, which was something of a snub.

This third and final leg of mass production was messy and inefficient. In a lean manufacturing improvement frenzy at Toyota in the 1950s and 60s, comparable to Ford forty and fifty years earlier, Toyota engineers attacked this problem. Their first great enhancement to the Ford system came in the form of a "pull system" through the use of a kanban, or signal card.

Just as with Ford, the popular misconception is that Taichi Ohno was strolling through an American grocery store and experienced a "Eureka" moment. Just as with the Ford assembly line myth, that simply was not the case. The first kanban was implemented at Toyota in a machining operation in 1953. Kanbans were adopted for ordering parts from outside suppliers in 1965. That twelve year gap was one of trial and error, idea generation and effort that went far beyond Taichi Ohno alone.

Ohno may well have originated the concept when he was working as the manager of machining operations in one of the Toyota plants. He wrote about actually devising the idea when he was engaged in what he called "thinking backwards." The final mechanism of kanban – the one that stuck – was the product of the grocery store inspiration. Before that, however, the notion of pull flow, as opposed to push, was being pursued in a variety of ways. Kiichiro Toyoda served a role much like that of Bob Galvin, CEO, at Motorola. He did not have anything to do with the development or implementation of kanban, but he was a big supporter of the idea of a pull approach, and he empowered Ohno and others to work the principle throughout the organization.

Concurrent with the kanban evolution was an integrally related concept called the andon, or a means of visual control. Later the andons became almost all colored lights, but originally they consisted of a motley collection of flags, lights and signs, all intended to enable shop management and engineers to see at a glance which operations were running as expected, and which were in trouble. With kanbans telling operators what to do next and in what quantities, and andons raising the alarm for any disruptions in flow, Toyota was able to replace the chasers, counters, and checkers with a simple, fast, virtually mistake-proof method of synchronization.

Kanban and andon became universal after Ohno was promoted to responsibility for overall production. From that position of influence, he urged and often demanded that production mangers, supervisors and engineers drive the concepts through the entire organization and, ultimately, through the supplier base.

> *Ohno and other Toyota managers are noted for telling people what to do but not how to do it. One day he told a group their job was to set the vision and to challenge people to find increasingly better ways to deliver the vision. Mr. Katsuaki Watanabe current president of Toyota, recently said, "I want Toyota to build a new automobile that will be able to drive from the West Coast of America to the East Coast on one tank of gasoline."*

The greater improvement at Toyota during these same incredibly fruitful days was a solution to Ford's basic business problem – adapting the incredible efficiency of continuous flow manufacturing to the reality of the market. Whether their problem was a small Japanese market, or whether they had the foresight to see that even a high volume market would demand a wide range of models and options really makes no difference. The only real flaw in Ford's system was its lack of flexibility and this had to be resolved.

Ford's famous statement, "Any customer can have a car painted any color he wants so long as it is black," an offhand comment made at a board meeting, does not reflect an indifference to the customer so much as a firm conviction that flow and cost are virtually one and the same. Variation in the product, in Ford's mind, inevitably meant slowing manufacturing flow in order to switch from one configuration to another. That slowing of the flow would drive the cost of everything up – not just the product with the different color paint or different size motor, or any other option being proposed to him. He was absolutely right.

Shigeo Shingo was fond of telling people that he taught Ohno everything he knew. Shingo may have been the teacher, but both knew that Ohno was the boss.

The big breakthroughs in machine set-up time reduction came as a result of Ohno continually challenging Shingo to cut times further, from an hour to ten minutes, then from ten minutes to four, then to two.

In fact, both men were so intelligent, and so committed to the concept of the Toyota Production System, that it is almost impossible to attribute the genesis of many of the principles to one man or the other. There is no record of any disagreement between them on any aspect of the system.

No mistake should be made concerning where the issue was at Ford with variation. The final assembly lines in Highland Park were plenty robust enough to accommodate variation and there were some options offered on the Model T. In fact, the 1915 Model T came in five different body styles – all black, of course. What could not stand variation were the machining and paint operations. Changing over from one configuration to another would have brought plant flow to a standstill.

Toyota's single biggest accomplishment, and what sets it apart from every other entity in manufacturing history, is that it found a way to reconcile Ford's fatal flaw with his great vision. Variation was not optional – the market demanded it. At the same time, variation disrupted flow and flow controlled cost. The management at Toyota in the late 1950s and early 60s conjured up Single Minute Exchange of Dies. Largely, but not entirely, through the work of Shigeo Shingo, set-up times in the machining areas went from hours to minutes. When this occurred, lean manufacturing was virtually complete. Ford's low cost, short cycle time manufacturing could produce as wide a range of products and options as the market wanted.

Shingo's contribution was primarily to give vision and direction to strategies begun at Toyota years before he arrived. Efforts to reduce set-up times began very early in Toyota history, but improvements came in small increments. They managed to get set-up times down from two or three hours to less than an hour. Then Shingo came in and conceived of and led the huge breakthroughs that took just about every set-up time to less than ten minutes. It took Toyota almost ten years to get all of their machining changeover times that low, the latter half and the greatest progress driven by Shingo. The actual, hands-on, creative work was performed by numbers of mostly young, hard working Toyota engineers. Those engineers were Shingo products.

I was at Dr. Shingo's house in Fujisawa, visiting the Doctor and his wife Umeko, when he received a gold painting of Mt. Fuji from Toyota. It was his "going away gift." They retired Dr. Shingo as their consultant near the time of his 80th birthday. It was a very sad moment for both of us. In my opinion, instead of a gold painting, Toyota should have given Dr. Shingo a check for a million dollars – and even that would not have repaid what he did for them.

Shingo also had an enormous influence on Toyota's progress in quality control – Ford's first fundamental. His work in poka yoke took the ideas and techniques of the quality gurus to a higher plane. He enjoyed tremendous success as a result of the simple notion that, if defects are physically impossible to begin with, no further quality control is necessary.

Shingo began his work with Toyota as a consultant in 1955 at around the same time the company was broadening its product line to go after their version of Sloan's "a car for every purse and purpose" concept. He maintained that relationship until a few years before his death in 1990. Soon after beginning to work for Toyota, Shingo left the Japan Management Association for the incredible reason that he thought they were charging Toyota and others too much for his services. He spent the rest of his career as an independent consultant, working for far less than the Japan Management Association had been paying him.

These were amazing periods in manufacturing history. Ford in the early part of the century began the push to lean manufacturing and pursued it relentlessly for more than twenty years without an example to follow. There were no textbooks or consultants. In fact, the paragons of the day were quite different – largely Frederick Taylor influenced manufacturers, focused on operations instead of process flow. Taylor acknowledged that he had little influence on Ford. He found it "most interesting" – and in all probability a sizeable bruise to his ego – to learn that Ford had leapfrogged the manufacturing world "without the aid of experts." Since Taylor and his associates were the only such 'experts' available at the time, what Taylor was really saying was that he was stunned to learn that great manufacturing improvement had occurred without the use of his brand of scientific management. He dismissed the Model T as "cheaply and roughly made," apparently unimpressed with the fact that Ford had sold $120 million worth of "cheaply and roughly made" cars in 1914.

In the middle part of the century, Toyota took up the baton and moved lean through its next significant steps. The Ford-to-Toyota handoff, again, was not as simple as a brief visit by Eiji Toyoda to the Ford plant in 1950, as some would have us think, nor was it as simple as Taichi Ohno reading Ford's book, *Today and Tomorrow*. In fact, there was a Ford assembly plant built in Yokohama in 1925. It and the Model T were well known to the Japanese. Just about everyone in the manufacturing world read Ford's book when it came out in 1926. The Toyoda family paid close attention to every aspect of the emerging automobile business from very early on.

Sakichi Toyoda was the consummate eccentric inventor, known to lie staring at the ceiling for days, until he would spring from his couch with an idea.
He visited America in 1911 after giving up on the automatic loom. Seeing the dismal state of American loom technology, he decided that his machine wasn't so bad after all and returned to Japan to complete it.
Just as important, while in the U.S. he saw his first Model T and knew he was looking at the future.

The original Toyoda – Sakichi – was an inventor, a genius, an eccentric, and a bit of a racist -- a man very much like Henry Ford in many respects. Both men lived life to the beat of their own drummer. Sakichi visited the U.S. in 1911, saw a Model T for the first time, and clearly saw the coming of the automobile age. The decision was made then and there that the Toyodas would eventually be in the car business – the automobile was simply too marvelous for an inventor and mechanical genius such as Sakichi to ignore.

Toyoda's claim to fame was the invention of the world's first truly capable automatic loom, the rights to which he eventually sold to a British firm for a half million dollars in 1929. He charged his son, Kiichiro, with the task of using the money to get the Toyoda family into the automobile business.

The Toyopet Crown

The car was a dismal loser in the United States, selling only 958 units. Many of them were put on a boat and returned to Japan so Toyota could study the situation and retrench. GM's top engineers laughed and called them junk.

In July of 2005, Toyota had its best month ever selling 216,417 vehicles in the United States. In the most recent J.D. Powers study Toyota Motor Corporation earns 10 of the top model segment awards, with the Lexus SC 430 honored as the highest-ranking model for the second consecutive year. It has been an amazing journey.

* * * * * *

I truly think the secret of Toyota's success lies in their two pillars: Lean – the total elimination of all non-value adding waste and Respect for People. When you invest in people as your most important asset, you get the biggest return. These are lessons still to be learned by most American companies.

Kiichiro Toyoda arrived in Detroit just a few days before the stock market crash of 1929 that triggered the Great Depression. He stayed for almost three months, much of it spent following Charles Sorenson around Ford's River Rouge plant. After his stint in Detroit, he went to England and France to observe the state of the automobile art in Europe. By the time the Trans-Siberia Railroad hauled him back home, he had a very good handle on automobile manufacturing in general, and Ford Motor Company manufacturing in particular

Again it was a sustained twenty year or more effort with no guidelines to follow. By the end of World War II, Ford had abandoned any pretense of lean and Toyota was making it up as they went along. Like Ford in the early days they were clearly following a goal and a set of principles, rather than a roadmap.

It is apparent that Ford, Sorenson, Harold Wills and the others of their era, followed by the Toyoda family, Ohno, and Shingo in their day were not simply technical geniuses who had a creative flash that instantly changed the art of manufacturing. These men (and others) were leaders and managers in the truest sense of those words. They had clear visions and put in place the management systems and philosophies that focused and energized large scale organizations with astounding results. They led the two greatest manufacturing enterprises in modern history.

While the changes at Ford were taking place, there were over 12,000 employees at Highland Park, six satellite plants, and production volumes of over 1,000 cars per day. Clearly nothing that happened in such a company was a one man show. The scale was even greater at Toyota. There were not enough hours in Taiichi Ohno's or Shigeo Shingo's lifetimes to personally implement all of the kanbans and quick set-ups that came to life in Toyota City. These leaders created the management systems and the environment that mobilized large numbers of bright managers, engineers, and supervisors to accomplish the great things happening in their companies. They defined the goals and invested heavily in training and communications to assure that everyone knew all of the techniques and tricks of the trade that would get them to shorter cycle time manufacturing.

> *Ohno picked a leadership group of 10 people, one from each of their top ten suppliers. These men worked fulltime under Ohno to convert their companies to JIT. Each month they would also meet with Dr. Shingo who continually challenged them to "Do it!"*

That lean manufacturing began at Ford in the early part of the century, and then continued its evolutionary path at Toyota is beyond dispute. Both companies built their manufacturing around the core concepts of quality, flow, and synchronization. It is amusing to read the number of 'experts' who will have us believe that Ford created something called mass production, and then Toyota came along, saw all of its flaws, and developed an alternative system called lean. No one should know any better than Taiichi Ohno and Shigeo Shingo the influence Ford had on Toyota. Ohno said, "I think that if the American king of cars [Henry Ford] were still alive, he would be headed in the same direction as Toyota." Shigeo Shingo put the matter more succinctly, if not more eloquently: "Toyota system is not a system opposed to Ford system, but rather enlarged and progressed the Ford system."

It is significant that Taichi Ohno and others at Toyota were quite familiar with Henry Ford's 1926 book, *Today and Tomorrow*. It was out of print and long forgotten, if anyone in the American manufacturing community ever knew of it in the first place. Productivity Press picked up the rights to it from Doubleday for a song after Norman Bodek heard of its influence and the prestige with which it was held in Japan. Only after the Japanese pointed it out did it become important in this country, with tens of thousands of copies purchased at the recommendation of Ohno.

This is not simply an interesting historical tale. This fact – that Toyota is truly a progression of Ford's manufacturing management system - is vitally important to our overall understanding of lean manufacturing. Understanding the progression from Ford through Toyota has a profound impact on the question of why lean manufacturing takes hold at some places, and does not in others. It gives us a crucial insight into the heart of lean manufacturing.

> *Toyota's patience and long-term vision led to their ultimate success. It reminds me of the story of the tortoise and the hare. America after World War II was light years ahead of Japan. While American industry "slept" at the side of the road Toyota just slowly plodded ahead. American industry supported a dichotomy with management on one side and workers on the other. Toyota drew on all of their strength, and demanded that everyone work as a team, use their brains, and grow.*

CHAPTER FIVE

The Sloan Model Takes Over America

"There's another trouble about theories: there's always a hole in them somewheres, sure, if you look close enough."
- Tom Sawyer Abroad ~**Mark Twain**

If we are going to understand what drove Ford, especially in the 1910s, and Toyota, especially in the 1950s, to make such enormous strides in advancing the manufacturing state of the art, it is only logical to look for their common factors. Continuous improvement was wildfire at Ford in the early days, as much or more than at Toyota later, and it was largely driven from the shop floor. This calls into question the supposition that national culture had much to do with it. It is hard to imagine two more opposing cultures than post-war Koromo, Japan, and turn of the century Detroit.

1910 Detroit was a raucous, wildly growing city turning the corner from being a Great Lakes trading crossroads to a center of all of the emerging mechanical inventions of the age. It was a multi-ethnic, multicultural boom town, filled with great builders, wretched scoundrels, and everything in between.

Koromo – soon to be renamed Toyota City – was a place with a long, deep, classically Asian tradition, now populated by folks who had quite literally been bombed into submission. The ancient culture was humble and recent dreams of dominating the world had been replaced with more modest goals of eking out a decent living. It was a city in a country faced with rebuilding, hardly one with "the sky's the limit" attitude of Detroit forty years previous.

The idea that there is something about lean manufacturing that requires a uniquely Japanese labor force or an Asian work perspective simply does not hold water. It was a tough, uneducated, not particularly subservient crowd in Highland Park that worked by an "Everyone takes the hardest way first" attitude; *kaizen* means about the same thing. While early Ford and later Toyota had similar internal, continuous improvement cultures, they came from radically different external, social cultures.

Leadership is another often-cited necessity for lean manufacturing success. Certainly that case can be made when comparing early Ford with Toyota. It stands to reason that any great organizational accomplishment requires effective leadership. Henry Ford and Kiichiro Toyoda were both men of vision and conviction, and they accomplished great things. At the same time, it must be pointed out that they were also both shown the doors to the companies they created, and there were few employees sad to see either of them go. It would be foolish to attempt to demean the leadership of either man, but they were terribly flawed in some aspects, to the degree that their leadership skills alone cannot be the reason for their accomplishments.

Kiichiro's kinsman and successor, Eiji Toyoda, was most certainly a visionary, an effective leader, and much more of a complete person than Ford or Kiichiro, but credit for his accomplishments must be tempered by the fact that he was able to stand on the shoulders of the giants preceding him.

There have been a number of effective leaders who have not accomplished anything of significance. For instance, General George Armstrong Custer had little difficulty getting people to follow him. As with every leader, it is clear that where the leader proposes to take his followers, how he plans to get there, and for what purpose have a great deal to do with things. Certainly leadership is essential to lean and it was a common link between early Ford and Toyota, but there is much more to it than that.

The more important common links between the two companies that drove them to make these amazing strides would appear to be the following: (1) they were privately and largely family owned, (2) they were extraordinarily single-purposed in pursuit of market share, and (3) they were driven almost exclusively by cash flow rather than any book measure of profitability.

These common factors provided the motivation to develop management systems and methods that inevitably resulted in lean manufacturing. This explanation also stands up when looking for the reasons for Ford's sudden and complete abandonment of lean manufacturing after World War II. When Henry Ford II took over, his family brought in a General Motors-experienced management team, and the market share/cash flow goals were replaced with the objective of maximizing shareholder ROI - Return On Investment. Almost immediately, lean manufacturing went out the window.

Alfred Sloan is probably given more credit than he deserves for modern management, but there is little doubt that he presided over the architecture of the management systems that exist in General Motors and Ford today.

A few other companies, including General Electric and International Harvester, claimed to have their own systems of management, but the differences between their systems and Sloan's boil down to nomenclature. The Sloan system is in place at just about every public corporation. That system is designed to achieve a markedly different set of goals than the systems at early Ford and Toyota.

Alfred Sloan was a serious man, as this photo shows. Nice enough, but somewhat humorless. He fought all his life to control the thick Brooklyn accent with which he spoke, reflecting his boyhood in New York City. A lifelong bachelor, General Motors was his life.

While the lean creators pursued cash flow and market share, Sloan companies pursued (and continue to pursue) Return On Investment (ROI). To be specific, they seek to maximize stockholder ROI.

There are many ways to get a return on an investment. For example, one person may invest in a house on the lake hoping to fix it up a little and sell it very soon at a sizeable profit. Another person may buy it hoping to use it every summer until his family is grown and then sell it at a price that allows him to get his money back out of it to pay for retirement elsewhere. His return is essentially free use of the home for a period of time. A third person may buy it with plans to use it as rental property, generating sufficient cash flow to make the mortgage payment, thereby having it largely paid for when it comes time to retire, then he will move into it. He is seeking basically to get a free retirement home. Three people – the same investment – three different ideas of return.

Pierre S. DuPont as he looked in 1920 shortly before taking over General Motors. His family owned what was initially a gunpowder business, and is now DuPont Chemical. Their $38 million bailout of Billy Durant put the DuPont family in charge of GM.

The point is that the Sloan system is not necessarily driven by the only view of ROI. It is driven by the definition of ROI that Alfred Sloan's boss – Pierre DuPont – had at the time Sloan was put in charge of General Motors.

DuPont's concept of return on investment from Dupont was measured by how much the corporation's shares were worth at any point in time, compared to how much the shareholders had paid for them. There is nothing inherently wrong with that, until one gets a little deeper into the worth of a share of stock.

The value of stock is basically the book value plus or minus what the owner of the stock and a potential buyer think will happen in the future. Book value is a very simple calculation: Add up the assets of the company, subtract the liabilities, and the difference is the equity, or worth of the company. Divide that by the number of shares of stock outstanding and the result is the book value per share. Temper that number by what management says the future will be, what a potential buyer thinks the future holds, what all of the other stockholders and potential stockholders are seeing in their own crystal balls, and the result is what that stock is worth on any given day on Wall Street.

As a result, in the definition of return that Pierre DuPont and others of his frame of reference used, the return ought to be what his share of the business was worth if the company were to be shut down and sold off tomorrow. It seems a rather curious way of looking at General Motors or General Electric, but that is how it was done. As a result, inventory, which to some looks like money spent unwisely, is an asset. After all, if the business closed tomorrow, the inventory could be auctioned off to someone, so it has value. People, who to most business owners seem to be very important, are of no value whatsoever in the DuPont model.

It certainly is one way of looking at things and it is the way virtually every big business does today, but it is far from how Henry Ford and the Toyoda family looked at their businesses. For one thing, neither owner had any intention of closing the business or selling it off to anyone, so such calculations would have been meaningless to either of them.

There is no indication that DuPont had imminent plans to sell his share of GM, but he apparently wanted to keep that option open at all times. He certainly expected GM to sell off any part of itself that was not pulling its weight. While DuPont may have been at GM for the long haul, the basic idea of the stock market is to buy shares in one of its listed companies, and then the minute it looks like the company's value on the auction block is at its peak, sell your shares. Most stockholders are not invested in a company for any planned period of time. They are perfectly willing to sell tomorrow if that is in their best interests, or a year from tomorrow, or ten years from tomorrow; and they expect management to be sure that the stock price tomorrow is as high as it can be whenever they choose to exercise that option.

The Sloan system of management is directed squarely at making sure that Mr. DuPont's idea about ROI drove the business. After all, Sloan was running General Motors mostly with DuPont's money, so doing things the way the boss wanted certainly made sense.

Since the first entrepreneur went out looking for capital, people have been painting an exceptionally rosy picture of the future and modern management is no different. To the extent that managers of these big companies can keep the stock prices up with plausible excuses for past failures and optimistic projections for the days to come, they naturally do so. But their primary tool for controlling ROI is the balance sheet. That base of the ROI calculation – assets minus liabilities – is, for all of the smoke and storytelling that goes on within the investment community, still the most important element, and keeping it 'strong' is management's primary task.

> *On one study mission to Japan, I led a group of top managers from Dana Corporation, including then chairman Joe Magliochetti. From earlier trips to Japan and meeting many Japanese managers, I knew the Japanese view was that ROI could be a very misleading measure. Without any deep thought about it, I told the group on a bus between plant visits that ROI was not a good measure and that they should abandon it. You would think that I just allowed the dam to break. Virtually every manager on the bus shot me down and made me feel like two cents. In one short moment I lost all credibility with them. The Japanese managers taught me that ROI was misleading because it leads you to focus only on the short term, to the long-term detriment of the company.*
>
> *A few months later, Woody Morcott, the next chairman of Dana, keynoted one of my Productivity "The American Way" conferences and told the audience that ROI was key, and the only important measure for his managers. I knew then why I had been so quickly shot down in Japan.*

The 'heavy hitters' on the GM management team. The painfully thin Sloan is on the left, Pierre DuPont stands in the center and the diminutive Raskob is on the right.
Raskob did not stick around for long. He committed the unforgivable sin of being a Democrat, which was one thing Pierre DuPont simply could not tolerate.

Sloan and others created elaborate systems and rules of management aimed at making sure the whole company was focused on keeping Mr. DuPont's number as high as it could be all of the time. The most perplexing aspect of the system, however, is the way it causes inventory to be managed, which in turn flows down and out with all sorts of ramifications on how factories are managed – most of them bad.

Since inventory is an asset and so is cash, they are virtually interchangeable. In fact, they both fall under the category of 'working capital,' further cementing the idea that they are pretty much one and the same. Whether the company has a million dollars in cash or spends it on a million dollars worth of parts really has little bearing, within a very broad reasonable range, on the net worth of the business. It has a million dollars worth of assets either way. As a result, within that very broad range, it appears as though having inventory is good if that inventory seems to help save money in manufacturing things.

The system even allows credit for half-built products. The inventory that was purchased can be partially built into a finished product, and then parked, and it is still an asset. In fact, all of the costs to half-build it are added to the original price paid for the material and carried on the books as an asset – presumably under the assumption that someone will pay the half-built cost at a going-out-of-business sale.

The philosophical underpinnings of much of this can be found in Sloan's "M Form" concept. One of his key ideas was to break General Motors into divisions that would operate 'independently' – decentralized, according to Sloan – but still reporting to and controlled by the corporate headquarters. Because it was really one company, this decentralization often required the divisions to 'buy' and 'sell' from each other. Fisher Body, for example, sold car bodies to the Chevrolet Division and so forth.

This notion of a company buying and selling from itself flowed down to the very details of the company's cost accounting scheme. Since the painting operation is not really a business and cannot actually buy anything, inventory was (and still is) the surrogate. Machining 'sells' parts to paint by putting them into inventory, where they will sit until the painting operation needs them. The idea is that somehow, even though nobody really sold anything to anybody, machining made a profit on the deal, and that profit can be used to measure the return on the company's investment in the machining department.

Sloan had less to do with this part of the system. The thinking here came mostly Donaldson Brown, who came to General Motors from DuPont to help out. Before he was through, Brown earned recognition in the financial community as the father of modern cost accounting. His reputation at GM was that of a sort of resident genius. He was so intelligent, many of his contemporaries reported, that after listening to one of his presentations, they often had absolutely no idea what he was talking about much of the time. Regardless, Alfred Sloan and Pierre DuPont understood him and that was all that mattered. (To be truly accurate from a historical standpoint, some credit should also be given to Jakob

Raskob, first a DuPont Treasurer then in the same role at GM and eventually the builder of the Empire State Building. He was something of a financial muse to all of them – Sloan, DuPont, and Brown.)

Brown's contribution was primarily that he could take this definition of ROI, look out over General Motors operations and envision islands of cost awash in a sea of assets. Those islands of cost were basically people. Well-supervised, they could turn one form of inventory into another with little of their time left over to detract from profits. Left uncontrolled, however, they could waste a lot of money with nothing to show for it. So 'responsibility accounting' came to be, where each of those islands was deemed to be a cost center, which added up to plants that added up to divisions that added up to corporate. In the end, every nickel spent could be traced up to Mr. DuPont's ROI number and everyone knew how they were impacting the price General Motors could command at a going-out-of-business sale.

Not only has this become the model for all of the big, publicly traded manufacturers, it is embedded in academia and in the law of the land. This logic and the systems that support it form the backbone of MBA programs. One who thoroughly masters it can become a CPA. A publicly traded company that fails to keep its books this way is likely to have to answer to the SEC and the IRS.

The people at Toyota and the early Ford management did not look at their businesses as though they should constantly prepare for a liquidation sale. They ran them as going concerns. In fact, they viewed them the same way as the average guy who owns a typical small machine shop does. He only looks at his balance sheet when he takes it to the bank to apply for a loan, or when he files a tax return. In between, he does not view it as much of a management tool. His definition of assets and liabilities depends on the circumstances. Looking out over his shop, if the material, people, and machines are all moving, they are assets. If they are standing still, they are liabilities.

CHAPTER SIX

Sloan, DuPont & Brown

"The significant problems we face cannot be solved at the same level of thinking we were at when we created them." **- Albert Einstein**

There is no question that Pierre DuPont, Alfred Sloan, and Donaldson Brown were extraordinary, intelligent men; and there is no doubt they knew money and finance as well or better than anyone in their day. In spite of that reservoir of intellect, however, it is hard to conceive of three people less qualified to devise and implement a system for managing and making the most of an automobile manufacturing company.

DuPont was born into a wealthy family that owned a huge gunpowder company near Philadelphia. During a brief family squabble, he left the gunpowder business and went off to Ohio to work as an executive in the Johnson Company, a steel business in which the DuPont family had a financial stake. His brief tenure in the steel business happened to follow immediately behind F.W. Taylor's engagement as a consultant with the company, as luck would have it. Throughout the job, Taylor was a regular correspondent with Pierre's cousin and partner, Coleman DuPont. After a few years, Pierre and Coleman DuPont returned to the gunpowder business when they and another cousin executed a sort of leveraged buyout. He was there in the top position when General Motors came along.

William Crapo "Billy" Durant

Durant was the flamboyant visionary who put GM together and nearly destroyed it – twice.

After his second and final ouster from GM, he opened a bowling alley in Flint, Michigan. Always the big thinker, he had plans to create a nationwide chain of bowling alleys when he suffered a stroke that ended his career.

Henry Ford once disparaged businesses that were big simply for the sake of being big. They did not serve any useful purpose, Ford asserted, they were simply "business museums." He was undoubtedly referring to the museum of businesses William "Billy" Durant was putting together that would eventually become General Motors.

The story of Durant and the origins of General Motors are an incredible tale in their own right. To sum it up briefly, however, Durant, part car man, part financier, and part salesman, pulled together a wide range of emerging automobile companies, mostly with IOU's and grand visions for the future. He built it up to a significant size, and then was kicked out as it began to come unraveled.

Incredibly, Durant then joined up with Chevrolet, built it up to the point that it was as big as General Motors, took over GM again, drove it to the brink of disaster again, and was kicked out again – this time for good. His final ouster coincided with, and was actually one of the terms of, the DuPont investment that bailed the company out. Pierre DuPont became the chairman, leaving his cousins behind to run the gunpowder business and the other DuPont holdings.

Among the businesses in Durant's original 'museum' was Hyatt Roller Bearing. The bright young head of Hyatt was Alfred Sloan. He was something of a prodigy at MIT, and had an engineering degree, as such degrees existed in those days. He had convinced his father to buy Hyatt. He then ran it as a profitable venture and eventually sold out at a handsome gain to General Motors, which he joined as part of the executive management team.

DuPont seated Sloan as the second in command and brought in Donaldson Brown to help run the finances. Brown was another engineer by education who had worked in sales for DuPont until he caught the eye of DuPont treasurer Jakob Raskob. Raskob made a finance man out of him, and DuPont plucked him from the fold to play a key part on the team that was charged with straightening out GM. Raskob preceded Brown in the exodus of DuPont management to General Motors, but was mostly the cash manager, while Brown and Sloan handled the operations.

The first order of business was to get rid of the museum pieces and put the whole operation on a paying basis, which Sloan and DuPont did quickly and well. The next task was to develop a means of managing and controlling the vast organization. That is where the trouble began that plagues manufacturing today.

On the scale of manufacturing complexity, steel and gunpowder dwell at the low end. That is not to say that they are not difficult endeavors. There are few environments tougher than a steel mill – especially back in the early part of the century – and few people tougher than steel workers. Gunpowder-making is not for the faint of heart either. In fact, not long before DuPont took over General Motors an explosion of epic magnitude at the DuPont mill in Fontanet, Indiana, changed the landscape permanently. Compared to automobiles, however, there is not much to steel and gunpowder. It takes a good scientist to get the recipe right, but it is batch processing and quality is pretty much a go or no go once the batch is brewed. Even with the machining operations typical in steel mills of the day, there were few steps. The steel was formed into a railroad rail, or a shaft, or whatever the principle product was, and that was it.

At the other extreme is automobile manufacturing. It encompassed over 5,000 or 6,000 components in cars of the day, hundreds of precision machining steps, dozens of sub-assembly processes, electrical assemblies, radiator construction, and then all of the sheet metal, painting, and cosmetic finishing. Getting all of this to come together is a challenge for anyone. With gunpowder and steel, there is not much flow and even less to synchronize.

Motorola's original Six Sigma concept centered on recognizing, mapping, and controlling "defect opportunities." Every critical dimension on every purchased component, and every time an operator or a machine touched a component, represented a defect opportunity. Automobile manufacturing presents hundreds of thousands of defect opportunities. Making steel or gunpowder presents a handful. The consequences of defects in steel or gunpowder formulation can be quite expensive and occasionally lethal, but the manufacturing management challenge to control them is trivial compared to a large-scale plant coping with millions of defect opportunities every day.

Three successful gunpowder and steel men assuming they knew the keys to optimizing the economics of an endeavor as complicated as automobile manufacturing is the equivalent of the winner of the America's Cup yacht race assuming he is qualified to command an aircraft carrier. Dupont, Sloan, and Brown were in way over their heads.

Their next order of business was the scheme to break the company up into divisions that focused on distinct products and markets – the "car for every purse and purpose" idea. While focusing the marketing efforts of the company in specific areas was a great idea, the company's divisions were actually still quite interdependent. The divisions 'bought' and 'sold' to each other, especially all of the component companies that were in divisions of their own. The idea of a company buying and selling to itself makes sense only on paper.

Donaldson Brown was schooled as an engineer, then went to work for DuPont in sales. John Raskob recognized his brilliant, analytical mind and moved him into finance.
He is known for his work in cost accounting, but before he created GM's flexible budgeting system, he took control over all purchasing and payrolls.
Under Brown, the only thing that was decentralized was accountability for mistakes.

Brown is reputed to have suffered from a severe drinking problem – a notion manufacturing managers would have no trouble believing.

This concept of divisions soon blasted down to the plants within the divisions and the departments within the plants, and eventually down into small groupings of machines within departments. With inventory declared to be an asset with the same liquidity as cash, it did not really matter whether the next 'cost center,' department, plant, or division actually needed the production output right away in order to consummate one of these paper sales. The producing department put the output into inventory and took credit.

Where Ford and Toyota worked to constantly tighten, streamline, and accelerate the manufacturing flow, DuPont, Sloan, and Brown's scheme was to break them apart. There is no question that this breaking the manufacturing company into modular components was intentional. Their idea was to have Sloan's old outfit, Hyatt Roller Bearing, for example, stand alone and 'sell' to the other divisions. Its 'profits' from these theoretical sales would be used to calculate the Return On Investment GM was realizing from its business. If it did not measure up, it would be sold off. General Motors was for sale in bits and pieces from the start.

That was problematic enough, but they went even further. Where their lack of manufacturing acumen really came to the forefront was in the financial controls largely attributed to Brown. He introduced the concept of flexible budgeting at the cost center level. The entire system is based on two terribly wrong and naïve ideas. First, Brown assumed that all manufacturing costs could be categorized as either fixed or variable. Second, the assumption was made that the variable costs varied with one simple driver. At first, units of production were used to 'flex' the budget up or down; eventually labor hours were determined to be the common denominator and they were used.

The idea that all costs behave in one of two simple modes – as a flat line regardless of volume or in a direct linear relationship to labor, or any other single factor, was an absurd over-simplification of the most complex business operation. While many costs may generally behave in such manners, very few truly behave so conveniently. General cost behavior may well be good enough at the top, corporate level where the last six zeros are often rounded off, but to drive these assumptions down to the factory floor, and then hold people accountable for actual costs compared to budgets based on such simple-minded assumptions was ludicrous.

They were deluding themselves into thinking that they had found a simple, logical, accurate formula to describe an incredibly complex system.

Further insight into the prevailing attitude at GM headquarters during all of this is Sloan's assertion that manufacturing was a secondary concern - that cars were basically commodities - and that sales and marketing were the keys to competitiveness. Manufacturing, in their minds, was a pedestrian sort of activity, to be controlled for sure, but not one worthy of much of their time and effort.

The net affect of the whole plan was to create hundreds of disconnected production organizations that produced only with bloated inventories. It was a scheme that would make quality improvement virtually impossible. It ran as counter to Ford and Toyota manufacturing theory as a business could run. Unfortunately, it seemed to work.

When the Sloan scheme of managing a manufacturing business was unrolled, the only real competitor was Ford and Ford was in a marketing free fall. The Model T had been in production for fifteen years when Sloan took over as president of GM. It was hurtling towards extinction, although Henry Ford could not see that. That Chevrolet outsold Ford soon after Sloan took over is hardly a credit to Sloan and it certainly is no validation of the system of management he devised. It is a clear statement of how far out of touch with the market Henry Ford was. Ford made a brief comeback with the Model A a few years later, but it was not the sort of product that could dominate for long.

By the late 1920s Ford was a rudderless organization as Henry Ford grew more bitter and cynical. General Motors took the lead in the automobile business and was never to relinquish it again.

CHAPTER SEVEN

The Collapse of Ford

"Adam was not alone in the Garden of Eden, however, and does not deserve all the credit; much is due to Eve, the first woman, and Satan, the first consultant."
- *Notebook, 1867* - **Mark Twain**

When a family with five or six growing boys sits down to dinner, there tends to be very little conversation. Speed is of the essence. Anyone who even slows down to season his food, let alone engage in small talk, will lose out. There is plenty of time later to talk, and even to consider whether the taste of the food was good. Japanese business operates on similar principles.

In spite of the calm and orderly outward appearance, and the near religious respect for civility and order, Japanese business is fiercely competitive. The formula for success that has evolved over the years is to go after the market aggressively, and worry about cost reduction later. The Japanese firm that waits while it meticulously analyzes every nickel of cost to be sure the margin is the highest possible, will find no customers waiting for them when they are finished.

The company that has a great deal of the market, with a break-even cost and price structure, is better off in the long term than the company with a great cost/price structure, but only a small slice of the market. The company with the larger market share will have a much easier time reducing its costs than the company with a small share will have trying to take customers away from the big guy. A "go big or don't go at all" mentality prevails.

As if that self-imposed pressure were not enough, Kiichiro Toyoda looked into the future after World War II and saw the American automobile companies coming. He made it clear to Toyota that they had three years, at best, to sew up the Japanese market. Otherwise Ford and General Motors would be there and it would be nearly impossible for an upstart Japanese company to break their grip.

Fittingly, when Toyota decided to go after the Japanese automobile market, they went after the whole thing as fast as they could. There were a few false starts. World War II was no small disruption to their plans. The initial entry into the automobile market was quickly snuffed out when the war began and the government dictated that they be in the military truck business. After the war, sorting out the new rules laid down by Douglas McArthur and the occupation forces took some time. Eventually they were back on track, however, and came out with the Toyopet Crown, a funny looking small car, but they were finally in the car business to stay.

If ever a company started out behind the eight ball, it was Toyota after the war. While Sakichi Toyoda left plenty of money and property behind, it was not quite as simple as funneling it into the car business and taking off. For one, the occupation powers were determined to knock the pre-war elite who controlled the money and industry down to size. They were also determined to put strong labor unions in place in order to put power in the hands of the common folk so that a few ruthless men would never again be in a position to cause trouble. Kiichiro Toyoda handled the new Japanese labor unions as ineptly as Henry Ford handled the UAW, which had a lot to do with his ouster from Toyota in 1950.

The biggest problem, however, was money. There was not much of it in post-war Japan, and what there was could only be found in the coffers of the national banks. The bankers were under the watchful eye of the occupation powers to be sure that it went toward efforts essential to rebuilding Japan's economy and infrastructure. The bankers were not easy people to work with even without their hands tied by McArthur's staff. While Taichi Ohno belittles the size of the Japanese market, stating that it was only a tenth the size of the U.S. market, which is quite an exaggeration, it was still significant. The company that set out to control it had to be prepared to spend better than a million dollars a day. For Toyota, that would come after they finished repairing the portion of their plant bombed in the last days of the war; and after they made a sizeable investment in capital equipment. All the while, they had to cope with bankers breathing down their necks looking for loan payments. In short, Toyota faced an uphill struggle for cash from the outset.

Just-In-Time did not spring forth from a burst of manufacturing genius. It came to be out of dire necessity. Toyota had trouble enough paying suppliers for the parts they actually needed. Paying for parts early would have been impossible. They made Just In Time arrangements for delivery because that was all they could afford. This same desperation for cash drove most of their manufacturing strategy, as we will see.

What made Henry Ford tick is still a matter of debate and speculation. But whatever his motivation, he was relentless in driving the cost of the Model T – and the price – as low as possible and shipping as many of them as possible. For his own reasons, his priority was also to capture the market.

Ford faced no such adversity as Toyota. There were plenty of banks and there was a flourishing stock market in the U.S. when the Ford Motor Company was formed and five years later when the Model T was conceived. Ford did not have to be self-sustaining for any practical reason. He simply refused to trust bankers or stockbrokers. As much as anything, he knew that other people's money meant less control and he was bound and determined to do things his way. Like so much about Henry Ford, the true motives for his actions went to the grave with him, so all we can do is guess.

While it was a dilemma entirely of his own making, Ford was in exactly the same position as Toyota would be later. He was obsessed with taking the whole market and he was compelled to do it all with cash flow.

A company that is driven by cash flow measures itself in a much different manner than one driven by DuPont's ROI. Henry Ford had a way of cutting to the chase, and he said that the only measure of performance that really mattered was the bank balance. If there is more money in the bank at the end of the week than there was at the beginning, it was a good week. If there is less, it was a bad week.

Even more telling was his statement: "Profit is the inevitable conclusion of work well done." Ford understood that the key to profits was managing the work, not the money. There is an outlook in cash flow driven companies that just about all costs are actually fixed. The responsibility accountants discretely break expenses down into variable, semi-variable or step costs, and fixed, based on an assumption that costs are under varying degrees of managerial control at all times. A more pragmatic manufacturing manager's view is that, while production is going on, the people, machines, and materials are already committed. There is nothing to be done about them once the game has begun. Financial results are much better controlled by managing flow, not managing costs.

Of course, cash flow is only optimized by assuring the entire production flow from the back door to the front door. No credit is given for partial work because, while Pierre DuPont may have known someone who was willing to pay for half products, none of Ford or Toyota's customers would. So Ford and Toyota measured themselves not on any incremental, operational cost level, but on the volume and speed of flow across the entire production process.

The managers at Toyota were a bit more sophisticated than those in the early Ford plants. They were certainly better educated. They discuss performance measurement in terms of cycle time. Far more eloquently than Ford, Ohno and Shingo explained that short cycle time can only happen with excellent quality. By its very definition, it drives down inventory and improves customer delivery performance and speed; and since it forces the elimination of any unnecessary activities, it assures low cost. Therefore, they point out, cycle time is the most significant measure of performance since it forces all of the important aspects of manufacturing to be optimal.

With the implementation of the Sloan system, General Motors transitioned from being a manufacturing company with marketing and finance functions to a marketing and finance company that happened to perform manufacturing functions. General Motors even maintained some executive and financial functions in New York City, instead of Detroit, making no secret of the fact that their focus was on Wall Street, rather than factories. Every Fortune 500 company followed suit, including many in moving the decision makers to New York City. In fact, some such as GE claim to have been well on their way to "M Form," ROI management before General Motors and DuPont's original company. Ford was the last true large-scale manufacturing focused company in the U.S.

Eleanor Clay Ford suffered the abuses her father-in-law, Henry Ford, heaped on her husband, Edsel, in silence until Edsel's death in 1943. When the old man moved to take the reins of the company again, she put her foot down, with the support of Henry Ford's wife Clara. From that day forward, Eleanor Ford was the undisputed leader of the Ford family, and she proved to be a very able matriarch.

On the fateful day in 1946 when Eleanor Ford called William Knudsen of General Motors looking for executive talent, the last chance for American manufacturing went out the window.

In 1943, Henry's only son, Edsel Ford, died. With more than a little justification, his wife Eleanor blamed Henry Ford. Her son, Henry Ford II, had been released from military service during the war to take over the company, but at 29 years old he was hardly qualified. As far as Eleanor was concerned, having old Henry return from retirement to help out was not an option. He was in his eighties, losing his health and angry at the world. She threatened to sell her stock unless he kept out. While she never had much involvement in the car business, Eleanor Ford was a smart, take-charge sort of woman and she emerged as the head of the family.

William Knudsen was an all-around capable manufacturing and business person with an unusually pleasant personality for an early car guy. For the most part, all of the men mentioned thus far were typical of the manufacturing men of their age – tough, serious, blunt, and fairly humorless. Knudsen was something of an exception, as Edsel Ford had been.

Bill Knudsen was working at a Buffalo stamping business – Keim Mills – when Ford bought the company. He was an excellent machine guy and a quick study. The machines and the staff all came to Detroit in 1913. Knudsen, however, although well-liked by just about everyone, did not feel the same about some of the Ford management – Pete Martin and Charles Sorenson, in particular. To accommodate him, Henry Ford put him in charge of building and running the remote Ford assembly plants.

To save on freight costs, Ford had erected a number of smaller plants around the country, especially on the west coast. Model T's would be built almost complete, but not entirely assembled. The engine and wheels still had to be mounted, and so forth. They found they could ship 26 automobiles in a freight car this way, as opposed to only 4 if they built them to final form. The assignment was ideal for Knudsen. He made a good living and made a substantial contribution. More im-

William Knudsen in the late 1920s, was, by all accounts, a very good man. But he was not at Highland Park during the boom years of the early 1910s and not the expert on the Ford / Lean manufacturing system many have made him out to be. He likely fell under the same misconception as the rest of America – that the Ford system was the assembly line. Ford turned inventory 40 or more times per year; under Knudsen and Sloan, GM turned inventory 6 times. While first Chevrolet, then all of GM, produced cars in staggering volumes under his management, the track record is one of producing at dubious quality and with steadily increasing costs.

portant to him, he stayed away from Detroit and the need to interact with Martin and Sorenson.

The problem with Knudsen, however, is that he played no part in the development of the Ford manufacturing process that was exploding at Highland Park while he was at the remote plants. That he was a sound manufacturing man was certain, and he had good business acumen, but he was not an expert in the Ford manufacturing system. At best, he was an arm's length observer.

In 1920, Knudsen left Ford on friendly enough terms and went to work for Chevrolet. He did a good job there and eventually rose to senior management at GM, but he in no way implemented the Ford system. Assembly lines were not the Ford system, they merely looked like Ford. In fact, just about everyone implemented assembly lines soon enough, and they did not need former Ford people to do it. Knudsen, Chevrolet and everyone else failed to implement the management system that drove the creation of assembly lines; and that was the system that drove almost zero defects and an intense focus on cycle time. Assembly lines without quality control and short total cycle times were not the Ford system.

It is important to note that the myth that assembly lines were the Ford system continues in the U.S. through today. However, for all of the credit Toyota gives to Ford for the origins of lean manufacturing, the writings of Shingo, Ohno and others make no mention of assembly lines whatsoever. When they say they learned it all from Ford, they were not talking about assembly lines.

After leaving Ford, Knudsen and his wife maintained a close social relationship with Edsel and Eleanor Ford. In 1946, General Motors was booming, while Ford was struggling. Knudsen was a friend, a former Ford guy, and a General Motors insider. It was the most natural thing in the world for Eleanor Ford to ask him for help in procuring a professional to come to Ford to help her son get the company in shape. Knudsen talked to Sloan who was eager to help. It was not that Sloan was particularly benevolent. He was concerned about the anti-trust laws and the consequences of having Ford fail. He knew that avoidance of monopoly charges demanded that GM have at least one somewhat worthy competitor, so he recommended Ernie Breech.

Ernie Breech was a cost accounting whiz and something of a General Motors trouble shooter. A GM division or plant that was not meeting its numbers could expect an extended visit from Breech that would last until the problem was straightened out or the under-performing operation was sold off. He struck Sloan as the ideal man to turn Ford around.

The day Ernie Breech walked into Ford was the day manufacturing died. He set about implementing the Sloan system, which killed the original focus on quality, flow, and synchronization. Within ten years Ford marketing and finance were strong, but quality was deplorable, inventories were bloated, and the factories were chaotic.

– Fortune Magazine described Ernie Breech as the man who would put the "GM organizational garment on the Ford manufacturing frame." No one knew at the time that doing so would kill lean manufacturing in the U.S. until Toyota brought it back with a vengeance 30 years later.

He came from the small town of Lebanon, Missouri and had a reputation for being devoutly religious off the job, and a tyrant on the job.

Henry Ford II put Breech out to pasture when he thought he had learned all he could from him. Breech then went on to head TWA.

Eleanor Ford, unintentionally to be sure, had thrown the baby out with the bath water. Ford was careening toward bankruptcy because of poor products. All of the manufacturing capability in the world is of no value if no one wants the product, and that was the case at Ford in 1946. She probably believed that relying on Knudsen for advice would protect Ford manufacturing, but he was the wrong man to ask. In fairness to her, there was no right man available to ask, and no one at the time could have seen through the fog to discern what the affect of hiring Breech would be. Even Fortune Magazine at the time described the move to hire Breech as putting "the GM organization garment on the Ford manufacturing frame." It was impossible to see at the time that the GM organization garment would crush the Ford manufacturing frame.

There is an often repeated accusation that Ford finances and accounting were in shambles when Breech arrived. What that really means is that Ford did not have a Donaldson Brown model cost accounting system. The lack of such a system was a problem only in the mind of Ernie Breech.

The example of financial disarray most often cited is that Ford did not have a handle on the exact number of outstanding invoices on hand in their accounts payable department, and that they estimated the amount by measuring the stack of unpaid bills when it came time to close the books at the end of the month. That accusation seems to be made exclusively by writers and academics with little experience in corporate accounting. It is unlikely that Ernie Breech was particularly bothered by that. No big company is in a hurry to process accounts payable, and estimating the amount by measuring the stack was a common practice among many Fortune 500 companies well into the 1980s. Not getting around to paying the bills promptly was more apt to be one thing Breech thought Ford did right.

The difference in approach to accounting and performance measurement was apparent in a discussion that took place at Ford shortly after Ernie Breech came in 1946. Soon after coming in, Breech was engaged in a discussion with a Ford manufacturing manager in which he was appalled that no one could tell him the cost to mount steering wheels. The Ford manager asked if Breech were suggesting that the steering wheel assembly was the bottleneck in the flow of the line. No, Breech answered. Perhaps Mr. Breech had noticed some inefficiency in the operation that they should look into. No, that was not the case either. That left the Ford man thoroughly confused.

They went around and around in the discussion. Breech could not conceive of how anyone could manage the assembly operation without a keen awareness of the cost of each step. He was worrying about whether the company was making a 'profit' from steering wheel assembly, and getting an adequate ROI. The production manager could not imagine what value knowing the cost of any particular operation was if it had no bearing on the flow and there was no reason to think it was being performed poorly. Breech won the bigger debate, of course, because he was the new boss.

Breech was engaged in pure Sloan thinking. The production manager was engaged in pure lean thinking. They were like ships passing in the night, but the effect of Breech taking control would have a profound impact on Ford and on American manufacturing.

Breech did not come in alone. He brought along a number of other accounting folks from General Motors. However, he was not the one to bring in the "Whiz Kids" – a group of statistical experts who had been influential in the War Department during WW II. They were more Henry Ford II's doing in hopes that their professional, logical, organized approach to war logistics would add value to Ford. For a while there was even a bit of a political rivalry between the Breech gang and the Whiz

Kids, but it did not last long. The differences between the two groups were minute. None of them were manufacturing people and none of them knew very much about cars. They all saw management as a grand exercise in numerical analysis and control, so they all ended up getting along well enough in the end. For the next twenty five years or more, virtually all of Ford's senior leadership came from either the Breech gang or the Whiz Kids, including Robert MacNamara, President of Ford, and the Secretary of Defense most responsible for the conduct of the Viet Nam War.

It was not Ernie Breech's aim to destroy Ford quality or to drive inventories up everywhere or to wreak havoc on factory efficiency. He was well intentioned enough. And it was certainly not the wish of the Ford family to bring these things about, but bring them about they did.

CHAPTER EIGHT

Factory Flow Grinds to a Halt

"Often the less there is to justify a traditional custom, the harder it is to get rid of it."
- *The Adventures of Tom Sawyer* - **Mark Twain**

Lean accounting – something along the order of how early Ford and then Toyota approached money – is not optional if a company wants to become a lean manufacturer. It is not just another area to work on, or something to pursue concurrently with kanbans and 5S programs. It is imperative and it is the key to the entire ball game. It is a basic prerequisite to lean manufacturing and, in fact, a company that converts to a lean accounting system will become a lean manufacturer whether it plans to initiate another lean implementation project or not. When accounting is changed, the rest of the business will inevitably follow.

Eli Goldratt, author of *The Goal* and originator of the Theory of Constraints, is a bit like Donaldson Brown in that there are times when he is so intelligent that it can be difficult to understand him, but any similarity to Brown ends there. Goldratt is crystal clear and speaking as plainly as possible when he points out that the goal of manufacturing is to make money. The importance of the accounting system is that it defines for the company what 'making money' means.

The Ernie Breech debate shows that this question is not as simple as it may sound at first blush. Breech, Donaldson Brown, and modern cost accounting have set in place a system that defines adding a steering wheel to a car as a money-making proposition, regardless of anything else that might happen to that car.

Credit for that work – it looks like a payment on the manufacturing budget – is given for performing that simple task because it moves money from expenses to assets. Partial, incremental work makes money according to Donaldson Brown and Ernie Breech. They seem to believe that manufacturing can meet Goldratt's definition of its goal – to make money - by performing any number of disjointed operations so long as it performs them to a standard without ever selling anything to anyone.

The lean accounting companies define making money as receiving cash. Incremental work is not good – it is a waste of money. Only completed work meets Goldratt's goal in a lean environment. Money is made only by selling to customers, not by transferring dollars between accounts or by moving parts from one inventory pile to another.

However the accounting system is set up, it defines 'making money' for the company and becomes the basis for all decision making. The quality system, production and inventory control systems and policies, people policies and so forth are all structured to enable the company to make the most money. How these systems are structured is a direct result of how the management of the company defines making money.

Traditional, Sloan type accounting begets traditional management and traditional manufacturing; lean accounting begets lean management and lean manufacturing. It is really that simple, and history proves it. Early Ford and later Toyota defined money in a lean accounting manner and lean practices resulted. General Motors defined making money as optimizing ROI and we all know the practices that arose as a result. Henry Ford and the Toyoda family did not personally go out and implement assembly lines or kanbans any more than Alfred Sloan went out and personally created batch production. These men defined 'profitability' for their companies, then urged and pushed their organizations to aggressively and creatively attain that version of profitability.

This view of management is hardly original. Hayes, Wheelwright and Clarke wrote about it in *Dynamic Manufacturing* in 1988. McKinsey and the National Center for Manufacturing Sciences both addressed it in white papers in the early 90s. They were all exploring how the best run manufacturing companies seemed to have a different and much more capable management infrastructure than traditional manufacturing companies. The systems, metrics, policies and procedures of lean companies all seemed to push the manufacturing floor to become leaner, while those same elements in traditional companies all seemed to fight the lean effort.

The historical evidence is compelling when we look at early Ford and Toyota. It is overwhelming when the post-war period and then the struggles with lean since the emergence of Toyota are examined.

The early part of the twentieth century was wild, chaotic, and exciting. It is interesting to study, but it was so helter skelter that it is difficult to draw many management lessons that are of value today, with the notable exception of Ford.

My grandmother, Ruth Waddell, was born in 1896 into a rural, horse drawn world of darkness, silence, and isolation. Twenty years later she went to the University of Wisconsin and into a world of automobiles, electric lights, radios, telephones, and airplanes. Never in history has there been a technological boom such as took place during her adolescent years. These product inventions sometimes led and sometimes followed equally important discoveries in materials and manufacturing – especially machining – technology. All of it shook out in a tumultuous process that determined which companies would succeed in bringing which inventions to the market and how they would do it. Some inventors and capitalists realized amazing fortunes, while most went bust. The confusion was further fueled by a world war, then falsely inflated by the rampant speculation of the 1920s, then brought back to earth with a resounding thud when the Depression hit in 1929. The wild roller coaster ride all came to a final stop with World War II. Success and failure, during those high energy years, proved to be more a function of invention and financing rather than management.

The effect of World War II was much like a grand game of musical chairs. The companies still in business when the music stopped on December 7, 1941 were not only given chairs, they were often given what amounted to a license to print money. General Motors, Ford, General Electric and the rest converted to war production and were kept busy, if not prosperous, for the next four years. When the war ended, they had vast, fully functional factories filled with machine tools. They also had plenty of cash, or at least a pocket full of government IOUs. More important, they also had the entire world market to themselves. The other emerging automobile makers, electric product innovators, consumer product companies, and machine tool builders of Europe and Asia were in ruins. Thirty years of grand delusion began. The harsh reality is that none of the big manufacturers, in spite of the billions of dollars they made, were ever particularly well managed. General Motors, General Electric, Ford, along with the new technology companies that emerged as a result of their war-time work – IBM, RCA and others – never made anything very well, except money. They simply had no competition other than each other. With all of Europe and most of Asia digging out of the rubble of the war, there were no alternatives to American products.

The arithmetic is fairly simple. Toyota committed to the car business in 1930. The war stopped them short. They recommitted to the automobile business after the war and got serious around 1950. By 1974, they were roaring down General Motors' and Ford's necks. The war set them back twenty years. Had it not been for that, Toyota would in all likelihood have been challenging the American automobile industry in 1954 instead of 1974. The vaunted Sloan system and its parallels at the rest of the big American industrials worked so long as they did not have to outperform anyone other than each other. Had General Motors been forced to contend with Toyota in 1954, the Sloan legacy would be much different.

It was around 1963 that I worked on a General Motors marketing survey. GM wanted to see how effective their advertising was. On one hand, they wanted to maintain their market share, probably 60 percent at the time. On the other hand they did not want to get any bigger and risk being perceived as a monopoly; so they were careful not to put Ford or Chrysler out of business.

The saying at the time was "As goes GM so goes the nation." We believed it without realizing that GM and the nation were going in the wrong direction, allowing European and Japanese companies to eventually catch us. In 1950, you could get 360 yen for one dollar. Today it is 112.65 yen to one dollar. Our great purchasing power was lost as we played with the numbers, instead of focusing on real productivity growth and quality improvement.

The management of General Electric for years thumped their chests and proudly let everyone in the world know that they had bested General Motors every time the two companies went head to head. The bragging came to an abrupt halt in the 1970s and 1980s, however. Then it seemed that GE could not go head to head with anyone, especially in consumer products.

General Motors, it turned out, was about the only company that GE could out-manufacture.

The notion of the 'professional manager' came into vogue not long before the American manufacturing house of cards tumbled in the 1970s. The theory was that a good manager could run any business. It was all done by the numbers, and knowledge of the product or the manufacturing process was not important. The theory was correct, because it was self-fulfilling. Every one of the big, public companies was managed exactly the same way, although they all spun their system with slightly different lingo. A man could go from GM to GE then on to US Steel and end up at NCR and not miss a beat. As it turns out he would not have been doing much good at any of them, but no one knew that at the time. They all managed by the numbers, and they all calculated the numbers the same way.

With all of the money being made, it was very easy to believe that we were, in fact, managerial geniuses. The American corporations ruled the world. The great business schools at Harvard and MIT – the Sloan School, in fact – put polish on the systems and cranked out people thoroughly steeped in the DuPont ROI way, and the money continued to roll in.

Along the way, the three fundamental pillars of manufacturing were knocked out from under these great companies (if they had ever really been there). Ford had pointed to quality, flow, and synchronization as the keys to making money in manufacturing. Almost by default, the first to go was flow.

> *I remember how proud Xerox was when it built what was probably the world's largest automated warehouse. Ohno would have laughed at them. To him, they were simply automating waste.*

By defining the creation of inventory, including work-in-process, as a money-making endeavor, any incentive to encourage flow went out the window. The 1950s saw the emergence of warehouses as a logical and necessary adjunct to manufacturing. Prior to that, the manufacturing warehouse was typically a relatively small shed out behind the plant. Warehouses were only found in shipping and distribution companies. By the 1960s warehouse space often equaled, or even exceeded, production space in many plants. The profession of Warehouse Manager had been added to the standard manufacturing staff.

In one of the greatest ironies in American business history, the Chrysler Corporation at one point had over 400,000 finished cars in finished goods inventory. Not only did they report a profit that year, they rented the abandoned Ford plant at Highland Park – the birthplace of lean manufacturing – to store many of them. One can only imagine Henry Ford spinning in his grave. One can also imagine Eiji Toyoda licking his chops knowing he was soon to attack the American competition. Yet the professional managers, confident in their mastery of the Sloan model and their money making prowess, recorded the figures in the 'good' column and collected their bonuses.

While the bloating of finished goods inventory was a visible sign of the lack of flow, it was hardly the only place where materials ground to a halt. Purchasing managers quickly figured out that buying several weeks, or even months or years, worth of something could be a money making deal. Since swapping cash for inventory had negligible impact on the balance sheet, saving a few cents by buying in bulk was simply smart business. Eventually all of that cheap material should come out the other end of the manufacturing process and then it would be a real savings. Inventories grew, flow slowed even more, and purchasing folks were given hefty raises for generating savings.

The most insidious deterrent to flow, however, was in the factories themselves. While Sloan, Brown and the rest may have looked out over the plants and seen islands of cost in a sea of assets, they knew that around the edges and lurking beneath the surface there were other costs in the form of overhead. The problem was that these costs, such as the costs of moving things around, fixing machines, inspecting parts, and supervising, were awfully hard to assign to a specific operation. They went along with everything in general, but nothing in particular. Without any means of directly assigning and controlling them, these costs were simply assigned percentages in the hope that they would stay reasonably in proportion to the direct labor costs which could be controlled, and that was good enough.

Without any direct link, all that could be measured with great confidence was the direct, easy to correlate part of the job: labor. It did not take much of a mathematician to figure out that, if all you really care about is the cost of performing one operation to a part, and you were allowed to make money by doing that single operation as cheaply as possible and then calling the partially complete product an asset, it would be cheaper to make them a bunch at a time.

It stood to reason that spreading set-up costs over many parts was cheaper than having to set-up for just a few even if it meant making more parts than you needed for a long time. It also made sense, if you could make enough parts all at once, to just make them cheaply, and then sort out the bad ones later.

Across the board, batches became the norm because the direct cost of batches was cheap and they could be immediately turned into money –

at least as far as Mr. DuPont was concerned – by classifying them as work-in-process inventory.

Once, I took Dr. Shingo to Yale University to deliver a lecture on quick changeover and the Toyota Production System to a group of students. After the lecture Dr. Shingo asked the instructor, "Are you going to continue to teach EOQ?" The professor was embarrassed, "Of course, I have to teach it. It is in the text books." Dr. Shingo just looked at him and laughed.

While Sloan and his minions may not have intended to drive inventories up, they were not unaware of them. In 1923, Ford's River Rouge plant was turning raw material into finished cars in four or five days. That same year, GM executive Albert Bradley presented a paper to the American Management Association describing the GM system that revealed their assumption that GM inventories would turn six times per year. Sloan and Brown clearly did not see that having flow through their plants more than ten times slower than Ford was cause for concern. They could hardly plead ignorance. In 1926 Ford boasted of cycle times at the Rouge in his book, *Today and Tomorrow*. Sloan knew what he and Brown had done to inventory. He simply did not understand manufacturing well enough to realize the destruction he had wrought.

Just as important, whether Sloan and Brown knew the impact of cycle time on manufacturing or not, and whether they intended to drive cycle times up or not, there is no doubt where responsibility lies. While Brown is heralded as the accountant who brought order to General Motors, it is not so well known that his first order of business upon taking control at GM was to assume authority for purchasing and production scheduling. In the General Motors factory, Brown decided what material came in and where that material went. Six annual inventory turns was the measure of his performance, but to a man who saw no distinction between cash and inventory, that performance was perfectly acceptable.

Note that Brown's next move was to assume corporate control over all payrolls. Unless a Brown underling by the name of John Lee Pratt signed off on it in advance, no one could be added to the company payrolls. With Brown in firm control of Men and Materials, all that remained was for him to implement the capital investment process so that he could control Machines. Then he had firm corporate control of all three "M's" of production, and manufacturing management had authority over almost nothing - just responsibility for the results. This, ironically, has come to be called Sloan's system of "decentralized" management. The only thing 'decentralized' is who would be fired if things went wrong.

This philosophy of selling to inventory became such an integral part of the means of making money that the numbers experts jumped in and made a science of it. While Ford and the folks at Toyota always assumed that the most economic quantity was one – or at least the quantity a customer wanted to buy right now – the professional numbers people knew otherwise and the notion of EOQ – Economic Order Quantity - became a fundamental part of the body of knowledge required to run a manufacturing business.

As the typical factory went from Ford's flowing river of material to a Sloanesque oozing swamp, those overhead costs crept up. So much focus on the direct costs created more and more work to handle, store, count and inspect the batches. It also knocked another pillar out from under manufacturing -- quality.

CHAPTER NINE

Manufacturing Quality Goes Out the Window

"If advertisers spent the same amount of money on improving their products as they do on advertising then they wouldn't have to advertise them." - **Will Rogers**

"The basic cause of sickness in American industry and resulting unemployment is failure to manage," said W. Edwards Deming. Deming, of course, was the quality preacher and statistician who found a much more attentive audience in Japan in the post-war years than he did in the U.S. He was brilliant and, at times, abrasive. The Japanese adored him. Along with others such as Joseph Juran, Kaoru Ishikawa and Genichi Taguchi, Deming found a manufacturing community in Japan that was eager to learn how to control quality.

The early quality leaders tried to point out the inherent flaws in batch manufacturing and inspection-based quality. It seemed perfectly clear to them that production in batches meant that it could be quite a while between the time a defect was created and the time it was discovered. When a week's production or more is made at one time, it would be a week or more until the next operation needed to use them. Any defective parts found at that step would be far removed from the event that caused the defect and it would be difficult, at best, to determine what went wrong. Long manufacturing cycle times make defect-free production virtually impossible.

In 1951, Dr. Deming, then professor of statistics at New York University, was invited to Japan by the American occupation command to teach quality control.. He taught a group of approximately 100 senior executives that quality improvement was a key to their nation's success.
At the time and for many years thereafter American managers were stuck with the mindset of the "military standard" plus or minus three percent – if we delivered 97 good parts out of 100 it was acceptable.
Dr. Deming taught them his famous "Deming Wheel": plan –do –check – act - meaning that quality can always be improved.

More problematic is the basic fact that no inspection is perfect. Some defective parts will get by even the most diligent inspector. That being the case, as more defects were found, the more defects surely flowed on to the customers. If the inspector is 90 percent efficient at catching defects and he discovers 90 bad parts, the math proves that 10 defective parts squeezed past him.

The only solution, declared Deming, Shingo, Juran and their contemporaries was quality at the source. That implied slowing production, or making more expensive machines and it was a tough sell in the U.S. The Japanese were all ears.

Some of the eagerness was the result of Japanese awareness of the reputation for dismal quality they had earned in the past. As hard as it may be to believe in the twenty-first century, there was a time when the words "Made In Japan" were synonymous with "junk." This became a matter of pride. The Japanese manufacturing community needed to prove to itself, as well as to the rest of the world, that they could make things as well as anyone else. While their defeat in World War II may have been inevitable, in any case, it was the result of many factors.

Certainly one major cause, and one the Japanese were well aware of, was that they lost to superior technology. Their aircraft, ships, and weapons were not as well-designed and not as well-made as those used by the Americans.

Most of all, the Japanese, in general, and the management of Toyota, in particular, were in desperate need of the economies that result from minimal defects. Kiichiro Toyoda said that if he were in "the sorry position of having to worry about quality, they might as well quit making cars."

When manufacturing is flowing one piece at a time – and a true assembly line is the epitome of one piece flow – a single defective part will cause the entire process to shut down. The cost is astronomical. The economy of continuous one piece flow is absolutely dependent on excellent quality.

Statistical Process Control – Deming's forte – appealed to flow manufacturers primarily because it is an effective tool for predicting defects before they occur. An SPC chart points to machines that are drifting away from perfection before they get to the point that the parts they are churning out are unusable. Heading defects off at the pass is essential to making money at flow manufacturing.

> *W. Edwards Deming – the "Father of Statistical Quality Control" - is greatly respected in Japan. While Shingo had a great deal of respect for Deming, he quipped that he was once completely convinced of the value of Statistical Quality, but after eighteen years of therapy, he recovered.*

Toyota went even further, however. Beyond SPC, they looked for ways to poka-yoke, or mistake proof, operations so that it would be physically impossible to create a defect in the first place. In this regard, they were imitating the early Ford quality system. Carl Emde's gang of designers was primarily engaged in conjuring up jigs and fixtures that could be added to machine tools in such a manner that they could be sure the only way to put a part into a machine was in the precise position to assure that the holes bored or metal turned was correct. Elimination of the potential for inevitable human errors to become defects was worth quite a bit of money and effort to these companies.

I remember the story told of a World War II Navy parachute factory. The Navy refused to accept the idea of Acceptable Quality Levels - they were not about to tell that to their pilots. To solve the problem, and to get zero defects, they told the assembly workers that they would have to jump out of an airplane to test the parachutes they made. Getting zero defect levels was not a problem after that.

When Dr. Deming finished his first lecture he was offered a financial honorarium. He refused to take any money and said that they should take the money and use it for a quality prize. His hosts, the Japanese Union of Scientists and Engineers, JUSE, took the money and created the Deming Prize. The prize became a driving force for companies to compete to improve their quality, and it worked.

Make no mistake that quality – or accuracy, as Henry Ford termed it – was economically driven, rather than customer-happiness driven. Toyota believed that customer satisfaction with the finished product was a function of design, more than production. The near obsession with quality at Ford and Toyota was propelled by the knowledge that absolute conformance to specifications enabled continuous flow, which was the lowest cost way to produce. Customer satisfaction with the fact that the cars worked as advertised was mere icing on the cake.

Shingo solved the quality problem through "self inspection" (workers are taught to double check their own work before passing it on to the next worker), "subsequent inspection" (the next worker immediately checks the part from the previous worker before they work on it) and the most powerful technique of all: poka-yoke *(asking every worker to come up with simple ideas for devices that would absolutely prevent defects from occurring).*

I remember visiting Komatsu and watching a worker use an automatic hand tool to tighten a bolt. The tool was calibrated to detect that precisely the right amount of torque was applied to the bolt; if not, an alarm would sound.

That philosophy must be contrasted with the modern American point of view. In the 1980s there were still 'professional managers' who were saying, "I don't know what quality is, but I know it when I see it." Henry Ford's definition – "accuracy" – was unknown to them.

Professional managers have long been big on acronyms. Along with ROI and EOQ was the concept of AQL – Acceptable Quality Level. It was a cost number derived mathematically by comparing the costs of inspecting, sorting, and knowingly shipping defective products to customers with the cost of improving quality. As expected, the heart of the arithmetic was the efficiency of saving labor cost through batch manufacturing. While the level of acceptable defects was essentially a comparison of the cost of warranties and customer returns with the cost of running an operation like that of Carl Emde and the cost of performing preventive maintenance to machines and tools, it was the batch that made this calculation possible.

When there is only one part available, the cost of it being defective is enormous – production slams to a halt. However, when there is a sizeable heap of parts to choose from, the cost of a defective one is minimal. Just toss it out, pull another part from the stack, and move on.

> *Learning is not compulsory... neither is survival.*
> -W. Edwards Deming

Of course, no one intentionally sent defective products to the customer. In actual practice, the numbers seemed to indicate that the least of all evils was inspection. That approach appeared to cost less than all of the mistake proofing engineering or the cost of returns and warranty claims. At any rate, a few irate customers and the cost of inspection were most certainly preferable to slowing down the pace of batch production that seemed to be the source of great efficiency and profits.

As AQL thinking took hold and batch manufacturing became the norm, the modern equivalents of Carl Emde – the manufacturing engineers – were directed more and more to focus only on direct labor savings. Machines that were more accurate or flexible could not be justified in a batch environment. Where Ford's engineers built machines to speed flow and assure quality, American engineers were pushed more and more into focusing exclusively on labor elimination technology.

The fact that defective products still seeped out of the system and customers of virtually all American made products came to expect defective stuff, was not really a major issue among senior management. Kiichiro Toyoda stated that he would rather shut down the company than be in the "sorry position" of having to worry about something so obviously vital as quality. At the same time, Henry Ford II, acting on the best information the dominating Ford finance department provided, pursued a strategy of setting the Ford dealers up with a repair system to fix all of the problems they could expect to find in new cars from the Ford plants.

What made such an incredibly blind attitude toward quality a bit more understandable was the fact that everyone was the same. Ford products were riddled with problems right off the line, but so was everything General Motors and Chrysler built.

> *During a visit to a GM plant, I watched a quality engineer check the tolerances on some stamped vehicle side doors with a gage. I saw a stack of doors on the floor and asked him, "How many doors have been stamped since this door was originally produced?"*
>
> *He answered, "Oh, maybe a thousand or so."*
>
> *I asked, "What would you do if you found a defect - the door is then out of tolerance; you then have a thousand other doors to correct?"*
>
> *He replied, "I would notify the die press people to make the necessary adjustments on future doors to correct the problem; then the new doors stamped would meet the exact specifications."*
>
> *"But what about the doors already stamped?" I asked.*
>
> *He seemed unconcerned. "They are normally not too bad, not that much out of sync. When they get to the assembly line the worst thing they might do is use a rubber hammer to make them fit properly into the car."*

A customer may not be happy with the fact that he needed a GE repairman soon after purchasing a GE television set, but that customer knew it would be no different had he purchased an RCA set.

The idea that all of these products were just too technical and too complicated to work right all of the time was so ingrained in the American consumer mindset that it never occurred to anyone that the root of the problem was just plain lousy manufacturing.

More important, the management that was driving the factories to the "sorry state" at which Kiichiro Toyoda would have been embarrassed to own them had the best of intentions. They were doing the best they knew how for their stockholders, customers, and employees. Their sin was not born of stupidity or negligence. Decades of ROI based management and Brown's responsibility cost accounting and flexible budgeting techniques had led everyone to believe that the cost numbers were right. The thinking prevailed that there was only one true cost for things, and the numbers they had were those true figures.

Their error was in believing the cost numbers. They were managing by the numbers and the numbers were driving them to that grand going out of business sale. Before too long, many of the companies would do just that. If not the entire company, at least many of the plants would be up for sale. Pierre DuPont devised a system to be sure that the salvage value of the companies in which he invested was high. From one end of the country to the other, GE and GM plants can be had for salvage value. The DuPont definition of making money worked as designed, but no one knew just how poorly it had been designed until the Japanese appeared on the horizon in 1970.

CHAPTER TEN

The Illusion of MRP

"I find that, as a rule, when a thing is a wonder to us it is not because of what we see in it, but because of what others have seen in it. We get almost all our wonders at second hand...By and by you sober down, and then you perceive that you have been drunk on the smell of somebody else's cork."

\- *Following the Equator* - **Mark Twain**

With the essential lean legs of flow and quality pulled out from under the factories, the final critical area of manufacturing execution, synchronization, had to bear all of the weight. With inventory everywhere and quality yields all over the map, synchronizing manufacturing so that anything could come out the other end became a difficult, but vital, task. IBM engineer Joe Orlicky, like the cavalry of old, rode to the rescue in the nick of time. In 1960 Orlicky installed the first MRP (Materials Requirement Planning) system at a J.I. Case plant in Moline, Illinois. The system was vastly complicated and took a computer the size of a small home to run, but it was a 'by the numbers' manager's dream come true.

Orlicky was a technical wizard, but suffered from the bland personality typical of IBM engineers of his day. MRP got its real boost from Oliver "Ollie" Wight – part expert, part consultant and part salesman who set out to create a career for himself and ended up creating an industry. Wight pitched MRP with the energy and fervor of a tent revival preacher and authored a 'must-read' book in its day called *Production and Inventory Management in the Computer Age.* He went on to energize the fledgling APICS (American Production and Inventory Control Society) into launching "The MRP Crusade" in 1971. It takes a couple of years of intense study and a demonstration of proficiency on a half dozen exams to become APICS certified, but that certification constitutes proof that the holder is competent in all of the intricacies of MRP.

Oliver "Ollie" Wight was the leading 'crusader' for MRP. Only 31 when he was creating the first MRP systems, along with IBM engineer Joe Orlicky and fellow consultant George Plossl, he led APICS and the mainstream of American manufacturing to MRP before his death at the too early age of 53.

Henry Ford accomplished synchronization for the manufacture of a thousand or more cars a day, with hundreds of complicated sub-manufacturing and sub-assembly operations, consuming more than 6,000 different parts with a Shortage Chaser, a blackboard and a handful of Checkers and Counters. Toyota did him one better, accomplishing the same thing with a deck of kanban cards and a box of andon light bulbs. In America we can no longer synchronize a factory that would fit into a corner of Highland Park without a quarter of a million dollar computer investment and a production control staff of certified MRP experts to run it.

What really gave MRP legitimacy was the conversion of MRP to MRP II in the 1980s. Again the acronyms can be overwhelming, but technically MRP stood for Material Requirements Planning, while MRP

The IBM RAMAC 305, which was the platform for the first MRP system in 1960. With five million characters of data storage, it was less capable than a modern video game system. Data was entered via punched cards and came out in the form of printouts. The hardware has certainly advanced since Orlicky's day, but the logic is almost the same as it was when he created MRP 45 years.

II stood for Manufacturing Resource Planning. That next generation of MRP effectively tied factory synchronization to cost accounting. In fact, it largely automated cost accounting, which made it nearly perfect in managerial eyes. Most important, with accounting in control of the important numbers in the MRP system, accounting was even more firmly in control of manufacturing. It has since expanded even deeper into the business and we now have ERP, or Enterprise Resource Planning.

MRP gave batch manufacturing an even greater validity than before. While MRP performs the necessary functions of 'exploding' a bill of material into all of its component parts and 'time phasing' the need for all of those parts over the life of the production plan, that is not really a difficult task. A fellow by the name of Fred Diehl accomplished it manually at Highland Park with the help of a few clerks. Toyota performed that function manually for years without a computer and without a hitch.

The complicated part, and the value of MRP, is its ability to sort through the mess that is expected on a factory floor, and plan for things to happen just as messily in the future. MRP assumes "batch and queue" production. That is to say, it assumes that parts will be bought and worked on in batch quantities that do not necessarily correspond too well to what customers actually want. It also assumes that this material will wait in line – in queues – for long periods of time between one operation and another. Waiting, of course, is an inevitable consequence of batch production. If the plant is going to produce several days or weeks of a part at a time, it is only logical that there will be a wait between batches. MRP also makes allowances for quality yields that are less than perfect and tries to straighten out mismatched parts sets. For example, it will see where the plant has one axle and three wheels and tell them how to re-align inventory to get the relationships right.

In short, MRP is based on the assumption that making things in quantities other than one piece flow is the way things are done, and it assumes that a lengthy cycle time from the arrival of material at the plant to its departure in the form of a finished product is normal. It also assumes that very little in the factory happens according to plan and that there are inventories at just about every level of production. With all of that as the expected state of a factory, MRP continually sorts it all out and re-plans production. The objective of MRP is primarily to drive the plant to an 'optimal' inventory level. That optimal inventory level is one that meets customer shipment needs at the greatest labor efficiency.

To describe the benefit of MRP more specifically, it allows each cost center to operate at its individual optimum without regard to the performance of the other cost centers. If the machining department is having a good week, that supervisor can claim credit for his production – perhaps even exceeding the schedule.

It does not affect him at all that the next department upstream – assembly, for example – is having major problems and will not come close to making schedule. MRP forgives this unbalanced situation. It and cost accounting will dutifully take all of the machined components and move them to inventory, cranking out reports indicating that machining did an outstanding job.

In 1980 I went to buy a Buick station wagon, but did not want any of the vehicles on the dealer's lot. I ordered the car I wanted – or at least thought I wanted -- from a catalogue. After 13 weeks my automobile arrived. A year later I visited Tokyo and saw a car dealer, but I did not see a lot full of new cars. If I wanted a new Toyota, I would have had to look at the catalogue and pick out the features I wanted. The difference was that, only one week later, the car would be delivered. It was my first realization that Toyota was doing something very different. A little over a year ago I ordered a computer from Dell with the features I desired – a few days later the computer arrived. We can do it here!

The early Ford and modern Toyota systems would have shut machining down at once until the assembly problems were solved. But MRP takes no such approach. It indicates that machining performed well to budget and schedule – never mind that they produced a lot of parts that cannot go anywhere. Then MRP will undergo one of its regular regeneration runs, consider all of the unbalanced and often unneeded inventories and a new forecast, and then re-plan everything in order to get it all back in balance. It never quite gets back into balance, however, because every time it runs some other department somewhere is over-producing and someone else is under-producing. MRP's primary task is to expect and forgive such sins.

MRP does this because its core is the logic and a set of algorithms to enable each component of a product to be produced at different volumes and speeds - and, in fact, the same components of a product going through different operations to be produced at different volumes and speeds - in order to optimize efficiency at each operation. It is based on the assumption that manufacturing is best performed in such a disjointed manner, and it assures adequate inventory to buffer all of this unbalanced production. Probably its most important feature is that it helps to make sure that every part bought and worked on is tracked into financial inventory. There is little doubt that Donaldson Brown would have liked MRP very much.

The lean literature is filled with theories and ideas about how to integrate Ford and Toyota type one-piece continuous flow with MRP, most of which takes a considerable stretch of the imagination. The way to reconcile lean with MRP is to turn the MRP system off. It institutionalizes batch manufacturing. Because batch manufacturing creates long lead times, MRP must operate to a schedule of future needs which are always unknown with any degree of precision. Inventory is the buffer between the reality of future demand and the forecast. Lean manufacturing, on the other hand, is reactive, responding quickly to actual demand regardless of schedules and forecasts. A lean "pull" system is essentially self-scheduling.

MRP is a system aimed directly at Pierre DuPont's ROI objectives, and it came to be because it made perfect sense to managers steeped in the ROI definition of making money. What made perfect sense to American manufacturing in 1960 is becoming more and more a monument to poor manufacturing. Much is made of the fact that MRP is a "push" system, while the Toyota Production System is based on "pull." This distinction is true and it is yet another weakness of MRP. However, the greater weakness of MRP is the basic premise upon which it is founded, which finds its roots in ROI-based financial thinking.

Taichi Ohno used the analogy of manufacturing and a river filled with rocks. Inventory is the water level, according to Ohno, that allows the ship of production to sail over the rocks. But the rocks represent wasted money and poor quality. Ignoring them by sailing over them is poor management. The goal of the Toyota Production System is to continually lower the water level in order to expose the rocks. That way they are recognized and eliminated.

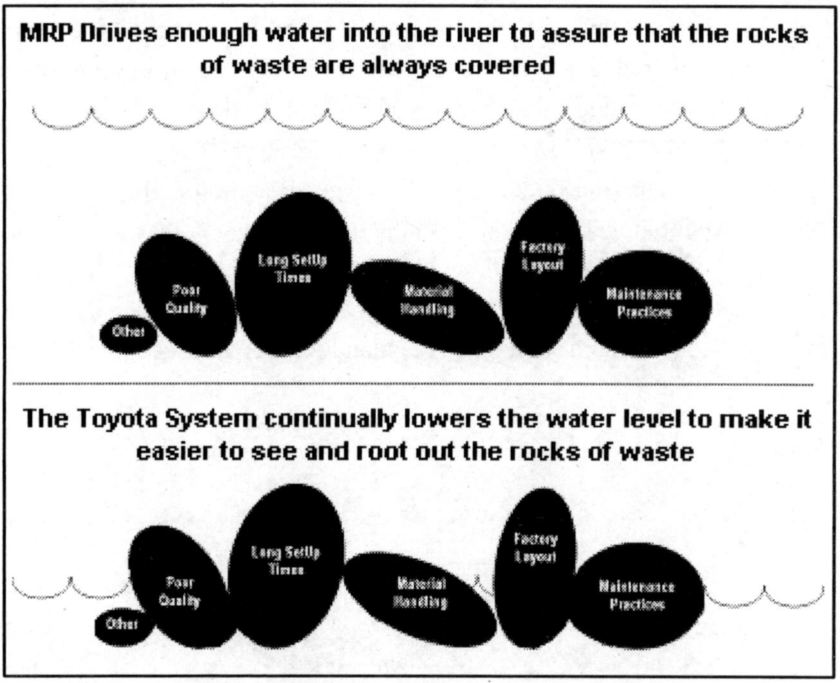

The basic logic of MRP is to input the location and size of all of the rocks – long set-ups, poor quality, unreliable machines – and MRP will work constantly to keep the water level just high enough to cover them up. Batch manufacturing – EOQ theory – recognizes a trivial cost of inventory, such as the interest rate money could be otherwise earning, and minimal storage costs. The EOQ calculation is essentially how big can the mountain of inventory be before the labor savings from batch manufacturing are all eaten up? When the formula is based on the notion that the only real downside of inventory is a few percentage points of lost interest income, the mountain can reach Himalayan proportions before it appears to be poor economics. The notion that the underlying problem that creates the need for batch production should be solved, saving inventory *and* labor, never enters the thought process.

MRP is based on an assumption that manufacturing is uncontrollable. It then helps drive so much work-in-process inventory into production that it proves that assumption correct. The Sloan/Brown idea of accounting shattered the first two legs of good manufacturing, and then built a vast, sophisticated third leg to rationalize the damage done.

The proof of the pudding is to engage in some of Taichi Ohno's "backward thinking." Would a manufacturing company that was operating on a philosophy of lead times of zero and lot sizes of one have any use whatsoever for an MRP system? The MRP crusaders would never have made it past the front door at Highland Park or Toyota City

CHAPTER ELEVEN

Blind to the Opportunities

"Leadership and learning are indispensable to each other."
From the speech he had prepared to give in Dallas on the day of his assassination. – **John F. Kennedy**

During the heyday of post World War II American manufacturing, management was a club for men only and every man wore a suit and tie to work, regardless of his level within the company. In Cincinnati, Ohio, the place for all of the managers, junior and senior, to buy the appropriate uniform was a chain of men's clothing stores called Burkhardt's. The fellow who did the buying for Burkhardt's waited to make his big decisions every year until the Proctor and Gamble annual report was released. Cincinnati is the headquarters location for P&G and Burkhardt's customer base included thousands of P&G managers. If the senior executives pictured in the report were wearing blue shirts or pinstripe suits, he tilted the stores' inventories to blue and pinstripes. If the executives wore white shirts or plaid suits, he stocked up on white or plaid. He knew that middle and lower level managers would follow the executive lead in lock step. The other huge concentration of manufacturing managers in town was at General Electric's sprawling aircraft engine facility, but those customers required no such market research. GE managers, without exception, year after year, wore white shirts.

Creativity had been so thoroughly driven from the big manufacturers that even the wrong color shirt rocked the boat. There was one way to dress, one way to manage, and one way to manufacture. Anything different was risky, and American companies had become incredibly risk averse. Risk was a product of engineering. These companies all had their roots in innovative technologies dreamed up by engineers in the early days, but Sloan had relegated technical people to second class status.

There are two paths leading to the employment office at the big manufacturers. One path meanders through business school, and the other through engineering school. When Sloan and Brown created modern American management, they gave all of the control to those who came in holding business degrees. Decision making was by the numbers – financial numbers – and the business school graduates were masters of the numbers. They owned the playing field and set the rules. Engineers seeking to try something different or innovative stood no chance. The discussion was never about technology. It was about "internal rates of return" and "discounted cash flow," all derivatives of DuPont's ROI. In Sloan companies the burden was never on the numbers people to understand the technologies around which the business existed. Rather, the onus was on the engineers to master the financial equations.

It is somewhat ironic that both Sloan and Brown were engineers by training, although neither of them actually worked in an engineering role. Sloan was a lifelong administrator and Brown had been a salesman before becoming an accountant. Perhaps they concocted a system that so stifled the technical people due to a fatal combination of theoretical knowledge without practical experience. They may have thought they knew more about engineering than they actually did.

A view of the Ford machine shop in 1913.

This was really the heart of the Ford manufacturing 'miracle' where millions of precision parts were cranked out daily in a continuous flow. Note the leather belts connecting the machines to the power shafts running the length of the shop. Later, the availability of independent electric motors to power each machine freed them from the straight line configuration and made U-shaped cells possible

In any event, manufacturing engineering was reduced to the exclusive pursuit of labor savings perceived to be attainable only through faster machines. Never mind that faster machines build inventory faster as well.

While much of the talk concerning lean manufacturing revolves around the elimination of non-value adding waste, lean manufacturers, in general, and Toyota, in particular, reduce direct labor costs at rates that leave Sloan companies in the dust. The critical technology to cutting direct labor hours by fifty percent or more is better than sixty years old. Electric motors small enough and powerful enough to drive a machine tool had a negligible impact on productivity in America, but a huge impact in Japan

When belt drives came off of machines, and each machine was powered by its own electric motor, the door opened up to a productivity improvement equal to that realized by Henry Ford with the advent of the assembly line. To companies viewing inventory as an asset, the productivity opportunity this opened up was very hard to see. With creative thinking all but driven out of the company, no one even went looking for it.

To look at an old photograph of a pre-World War II machine shop was to see a forest of leather belts, one arising from each machine, looping around a long metal shaft running the length of the shop. Early machines had no independent source of power. Machines were driven by the belts, which were driven by the long shafts, which were driven by power plants, usually steam, located just outside the factory. If Henry Ford did not do much tinkering with creating the assembly lines, it was due in part to the fact that he spent a lot of his time tinkering with the huge power plant that drove the machines at Highland Park. He loved the Model T, but he loved power generation technology just as much. One has only to look at the gleaming power plant he built to run things at his mansion to see where his heart and mind were.

One offshoot of this system was that machines had to be aligned in long rows. The turning shafts were a hundred feet or more long, with machines lined up directly beneath the shaft. A man could possibly run two machines at once, but even that was difficult. To run three or more, a man would have to be on roller skates, racing up and down the row of machines. For the most part, however, the arrangement was one man to one machine. While the power from the belt turned the cutting tool, it still was largely a manual operation before the war, requiring full-time attention to feeding material to the cutting tool, or engaging the tool once it started spinning.

To give Frederick Taylor his due, it was this belt drive technology that formed the basis of his early work and he did some pretty good engineering work when it came to belts. Prior to his day, engineers were expected to focus only on technology, without regard to dollars and cents. Such base concerns as money were seen as beneath the dignity of a technical expert. He conducted a detailed study of the cost of very sturdy leather belts that would last a long time versus cheaper belts that would require more frequent repair and replacement. The expensive, sturdy belts were the better deal, according to Taylor.

Taylor aside, the day came in the evolution of electrical technology when each machine could be equipped with its own motor. Motors were powerful enough, small enough and cheap enough for the shafts and belts to go by the wayside, and Taylor's study became moot. Without dependence on the shafts, machines could be placed anywhere.

To American thinking, this was not much of an event. Sloan's system was firmly entrenched by the time the shafts and belts were eliminated. Economy was perceived to result exclusively from running machines as fast as possible, making big batches at a time. There was still one man to one machine, for the most part, and maximizing the output from that man's labor cost was the objective. Whether machines were lined up in rows, or scattered at random around the factory did not make much difference to the results of that equation.

Shigeo Shingo presented a paper at a technical conference conducted by the Japan Management Association in 1946 entitled "Production Mechanism of Process and Operation." It was based on the principle that optimizing the overall production process – the complete sequence of operations that take a product from raw material to completion – is the key to manufacturing. To quote Shingo, "Improvement of process must be accomplished prior to improvement of operation." While the Americans saw manufacturing as a set of isolated operations, all linked by sizeable inventories, the Japanese saw manufacturing as a flow. Where the machines are is a big deal to people concerned about flow, while it matters little to people concerned only with isolated operations.

To Shingo, the flexibility to put machines anywhere he wanted opened the door to fantastic productivity improvements.

To a thinker like Taylor, and all of his modern American descendents, if a machine can produce a part every thirty seconds, it must be run at that rate in order to be efficient. That means that an operator must be there every thirty seconds to pull a completed part out and feed another part to the machine. If one believes in the principle that the lowest labor cost can only be obtained from running each machine at the fastest speed possible, then there are really only two ways to improve the efficiency, or productivity, of workers. One way would be to figure out a way to make the machine run faster. If it can be made to produce a part every twenty seconds, for example, the operator can produce a few hundred parts much faster than he could at thirty seconds each. The other possibility is the Holy Grail of American productivity improvements. If the machine can be fully automated – say, with a robot – then the operator would not be needed at all.

Shingo and Ohno saw a third way that was only visible by challenging the heart of the American economic assumption. They came to the conclusion that, just because a machine and a person can produce a part in thirty seconds, it is not necessarily economical to do so.

The word 'takt' is the German word for the baton that an orchestra conductor uses to regulate the speed, beat or timing at which musicians play. The Japanese took it to mean something like 'pace' and many Americans took it from the Japanese and are still scratching their heads wondering what it means at all. To Toyota it means the time interval necessary to meet customer demand. If the customers want 100 per day, and the plant works 500 minutes per day, then the takt is five minutes. Takt is the number that creates a perfectly level flow through the factory that meets customer demand.

To Shingo's way of thinking, making a part every thirty seconds simply because a machine was capable of doing so was not necessary if the demand was for a part every five minutes. In fact, running a part every thirty seconds was a problem because it wedded the worker to the machine. If that machine were to produce a part every five minutes – exactly as demand required – that worker would have four and a half minutes between parts to do something else. That 'something else' was to run another machine. In fact, by bringing all of the machines in a process together, that worker would probably still have time left over to run a third machine, too, or maybe even a fourth.

The key was not to have a random group of machines clustered together, but to have all of the machines in the same process with the same takt together. When the shafts and belts were replaced with individual electric motors, machines were placed in the shape of a 'U' and the possibility opened up for one worker to be no more than a step or two from several machines. The fact that the machines ran at different speeds was fine, so long as all of them ran at or below takt.

In one fell swoop, the number of workers required to run several machines dropped by a huge number. Not only was an enormous reduction in direct labor possible, but the machines needed were cheaper. While American engineers were driven to find machines that could reduce the time to make a part from thirty seconds to twenty seconds – which invariably meant a more expensive machine – it was OK with Shingo to replace the thirty second machine with one that could only produce a part in sixty seconds – a cheaper machine.

To put the last dollop of icing on the cake, running everything at takt meant that parts could be made one at a time. One piece flow, rather than the batches that resulted from the high-speed operational focus, meant less inventory. Less inventory meant less material handling and less floor space.

Stubbornly clinging to Taylor and Brown's principles, the big American manufacturers still chase the original two options believed to be open to them. When driving a machine to faster speeds is no longer feasible, the alternative is to find a cheaper worker. However, one person working on one machine, going hell bent for leather to produce as fast as possible, still cannot compete with one American or Japanese person working in the middle of a cell running four, five, or even more machines, even if that one person on his one machine is working for Chinese or Malaysian wages. While American companies outsource, Toyota puts machines on wheels so it can keep moving them around to reduce labor even more.

We are going to China to save labor costs when one-piece-flow, properly implemented, will cut labor costs drastically. Recently, I visited a plant where five people were standing in what looked like a one-piece-manufacturing cell, But, the workers stood in front of their machines waiting for a part to be handed to them. If the workers were multi-skilled and simply walked along the cell, they easily could have reduced the number of people from five to two without having to go to China to save labor. Toyota discovered that focusing on improving labor productivity, within the constraint of takt, was much more advantageous than focusing on machine efficiency.

The second option, eliminating labor all together through automation, is still an objective that draws American managers like moths to a flame. Toyota, however, sees pursuit of full automation as a fool's errand, for the most part. Just about every automation project ends up still requiring people. Perhaps they no longer run the machine, but they are needed to monitor it, move material to or from it, or keep tabs on the quality. For whatever reason, there is generally still a labor component after the expensive automation investment is made. They have learned that semi-automation with one person in the middle of several machines gets the same result with far less expensive machines.

The ability to see such a staggering opportunity to reduce labor content from something as bland as swapping belt drives for electric motors requires a fair amount of initiative and creative thinking, as well as a strong sense of the integrated nature of manufacturing. It is far beyond that which can be expected from people who must worry about being the only one at the meeting with a blue shirt. In most companies, any engineer who dared to challenge the status quo and the unquestionable truth behind the accounting principles with a proposal such as Shingo's would certainly have faced rejection. In some companies, such thinking would have branded the person making such a suggestion as a sort of an incompetent trouble maker. Sadly, the state of American manufacturing had become such that it was incumbent upon the technical people to conform to accounting principles, rather than have the accounting people adapt to evolving technologies.

CHAPTER TWELVE

End of the Quarter Shuffle

"Figures often beguile me, particularly when I have the arranging of them myself; in which case the remark attributed to Disraeli would often apply with justice and force: "There are three kinds of lies: lies, damned lies and statistics."

\- *Autobiography of Mark Twain* - **Mark Twain**

Senior managers, business school professors and strategy consultants are rare figures in factories at all, but they can never be found in one during the last few days of a fiscal quarter. There are precious few Wall Street analysts around either. If they were to visit the factories of just about any publicly traded company during those chaotic few days, they would see just how powerfully destructive – and diametrically opposite to lean theory – their ideas about manufacturing management truly are.

During those last few days of the quarter, all of the stops are pulled out to make the numbers. Some of what is done is downright illegal. Much of it is unethical. Just about all of it is senseless. However, management salaries and promotions are driven by numbers. Plants are closed or kept open based on the numbers. Work is shifted from America to China and India based on these numbers. The numbers, unfortunately, are not based on sensible economics or what is good for the long term health of the business and its various stakeholders. They are based on Wall Street schedules and balance sheet targets.

Most plants have been through the drill so many times that the people need very little coaching. All the way down to the operators on the production floor, the people know how to rig the numbers.

Every machine and every production employee will run all-out if there is anything in the plant for them to work on. Supervisors who are normally lenient concerning breaks will become frenzied taskmasters. Of course, there will be no employee safety or quality meetings. Nothing will keep production people from their machines on the last few work days of the quarter.

There will be a steady stream of shop supervisors to the production control office demanding that production orders be written to machine, assemble, paint, or pack whatever parts they have been able to find in the plant. If there were a demand for those parts they would already have production orders, but they are not producing for demand. They are producing for the sole purpose of earning credit for direct labor hours, which, in turn, earns credit for overhead on their budgets.

Quality inspection will virtually shut down. If anything produced is bad, no one wants to know about it until next week – after the books are closed on this quarter.

The folks who communicate with customers will be on the telephones trying to pull a few orders for next month into this month. The fact that this strategy rarely works – the customer is often going through the same quarter end numbers game – really does not matter. The shipments will happen anyway. At least they will on paper and, so far as accounting and Wall Street are concerned, paper is all that matters.

Every shipping and receiving manager and employee worth his or her salt is a master of the game. Product not scheduled to ship for days or even weeks will be pulled, skidded, wrapped, labeled and entered into the system as gone – then stacked off to the side. It will not really go anywhere until closer to the true customer due date.

With work-in-process inventory building in the plant at staggering rates, the plant inventory goal may be in jeopardy, so receiving virtually shuts down. Trucks are turned away when possible. If they have to be unloaded, their contents will be stacked on the dock, but not received into the system until the next week.

The plant manager will spend the day sequestered in his office tracking the figures while a steady stream of production and supply chain people bring him the latest updates. He can be found behind an overflowing in-basket, because plant spending was shut down days before. Requisitions for maintenance supplies, training materials, and anything else deemed not critical to production will not be approved until the next quarter.

While the frenzy on the shop floor and most of the tension in the front office die down after the close of the quarter, the numbers game continues. There is a gap between the last day of the quarter and the final closing of the books, and in that gap, creative thinking is at a premium. If the maintenance expense was over budget, there is certainly some repair that was done somewhere that can be construed as adding to the life of the machine, so it should have been capitalized and written off over time, rather than expensed with the full brunt falling in the quarter.

Somewhere else, there most certainly was an overstatement of the amount of defective material to be scrapped. The plant decides they were too hasty and puts the material back into inventory so engineering

> *We must counter the enormous pressure placed on managers to meet or exceed the numbers. Wall Street demands that companies beat their projections and the penalty for failure can be very painful. Stock prices drop drastically when profits are off by only a few pennies per share. While executives may want lean, they must answer to Wall Street. Wall Street wants profits now, and higher inventory and slower write-offs can yield short term profits. This might be the primary reason why very few public companies, outside of the Toyota group, have successfully become lean.*
>
> *When a plant has a Kaizen Blitz, and makes substantial improvements in cycle time, the short term financial numbers can get clobbered. Converting inventory to cash makes book profit look worse.*
>
> *When we ran the first Kaizen Blitz at Jake Brake and moved 50 machines in one night, it was clear that this could be a real shot in the arm for American manufacturing. I wish I knew how to patent the event. 50 participants, under the tutelage of the Toyota masters, calculated both cycle time and takt time. They arranged the machines into five product lines, taught the workers and managers how to work in the new lines, and listed hundreds of small problems to address in the next few months. Despite some temporary upheaval, in a few months Jake Brake was way ahead and Danaher, their owner, was so thrilled that we were asked to run Kaizen events in all Danaher plants. Since then, the same exercise has been run thousands of times all over the world.*
>
> *It is short-term thinking when we focus on profits this quarter and go crazy building inventory to meet the numbers; but it is understandable when people's jobs are on the line. How do we change this? A few companies have found a way and the long-term results have been superb. Their senior managers understand that they are in it for the long term, and are willing to change their measures for success.*

can take a look at it at some future date to see if there isn't some way to make use of it. On and on the games continue until the numbers are right, or at least as right as they can get.

This scenario, to varying degrees, happens at every plant every quarter. When the plant and the company are having a good quarter –

'good' meaning high volume production and shipments with minimum labor – the insanity level is less, but still there. At the end of a bad quarter, it can be ugly.

Quarter end is the extreme situation, but the extreme serves to best illustrate the impact of the accounting system on manufacturing management. Manufacturing people are not the dolts the corporate staff people often make them out to be. They have figured out the game and play it well. While the average shop foreman may not know that the debits go on the left and the credits go on the right, he often knows better than the cost accountant exactly how production and labor reporting into the systems affect the reports used by senior management to gauge his performance.

Just about every plant has a meeting of some sort every day involving plant management and all of the people with some degree of influence over production. Those meetings center on two numbers: production quantity and number of direct labor hours. The previous day's numbers are always the primary topic. How much did we make and how many people did it take to do it?

The plant manager sets the tone for the degree of civility. Sometimes these meetings are supportive, with management attempting to assist the shop floor people in their efforts to improve these two numbers. In other plants, the foremen refer to them as daily "beating meetings." In some other plants, management has very little comment. They make the point with the sheer volume of detail discussed. Usually there are bigger weekly or monthly meetings that go into these two numbers in more depth.

The tone or the nature of the meetings really does not matter. The significance is the relentless nature of the meetings. There may be sporadic or less frequent meetings to discuss issues such as safety, quality, the upcoming employee picnic or the company's lean initiative. Every day, however, there will be a review of production volume versus labor hours.

> *Not too long ago, I visited a company and saw millions of dollars of obsolete inventory collecting dust on the shelves. I understood why the inventory was still there – to avoid having to reduce profits this quarter by writing it off. What I didn't understand was, when would the appropriate time be to "bite the bullet" and write it off?*

There is no doubt in the minds of any of the plant staff where the priority lies. In one major Midwestern Fortune 500 manufacturer, there is a meeting every morning involving virtually everyone, other than the hourly employees. Attendance is mandatory. The previous day's production volumes and headcount are reviewed for every product and every manufacturing operation. On a good day, the numbers are near expectations and the meeting ends within forty-five mind numbing minutes or so. On a bad day, some numbers come up short and the entire staff will suffer through a professional, but intense, grilling of the people responsible for the bad numbers. Following that meeting, the plant manager will participate via teleconference in a similar meeting with all of the other plant managers, and go over the same production and headcount details with the corporate staff.

Incredibly, this company has been engaged in a "lean transformation" for the last year. Their CEO has a resume that touts him as a man of considerable lean competence. Because the daily production meeting calls everyone together, the folks involved in lean usually hold their lean team meetings immediately following the production volume meeting. After being sent the relentless, formal message that earning standard labor hours is the primary goal of the plant. these people then break out to figure out how to comply with the corporate lean mandate.

While it may appear as though the lean team is being asked to fit a square peg into a round hole, they handle it adeptly. On the one hand, lean dictates that they drive the plant to short cycle time, high quality production with no waste, precisely leveled with customer demand. On the other hand, the daily meeting reminds them that the objective is to utilize labor without regard to inventory and marginal regard for quality. The dilemma is resolved by creating a plant that looks lean, but runs as it always has.

A U-shaped cell here and a kanban there is all it takes. When senior management comes to visit, bringing along customers or Wall Street analysts, they are shown visible proof that the plant is, indeed, lean. In the meantime, the plant continues to crank out volume. Of course, the plant and the company are not lean and sooner or later having only a lean façade crumbles. The company is typical. It is one of the 98percent that claim lean accomplishments in public announcements, but bring none of those accomplishments to the bottom line.

If you were to speak to the CEO he would undoubtedly tell you that his responsibility is to exert leadership, and he does that by talking about lean, authorizing spending on lean training, and bringing in lean consultants. The actual change to lean, however, is viewed as purely a factory issue at the end of the day. If you were to ask him exactly what has changed in his world, or anywhere at the corporate level, the truthful answer would be, "Nothing." Lean at the corporate level of this and virtually every big company is a factory philosophy, rather than a management philosophy.

This company has joined the wave of companies pursuing lean by launching a Value Stream Mapping exercise at the 'Enterprise' level. Participation in this effort would be the answer you can actually expect if you were to ask the corporate folks about their personal contribution to lean manufacturing. Since the corporate staff is not part of the value stream, they seem to be exempt from the analysis. They do what they have always done – analyze the rest of the business to see what lower level groups like the plants are doing wrong. What comes out is an analysis by corporate folks of what the plants and supply chain people need to do to become leaner. Nowhere in the equation does it occur to them that the plants and supply chain have come to exist and operate in a manner that best serves the definition of good performance laid out by the accounting system. The lean strategy they push down directs the factories to operate in a completely different, lean manner than in the manner which has proven to best meet the numbers; but they must still hit all of the numbers.

Without changing the accounting system, this is the inevitable result. The appearance and trappings of lean will crop up from place to place, but the company will not become lean. More often than not, when the lean pressure eases from corporate, even those trappings will disappear. The U-shaped cell will be straightened out to make room for more work-in-process inventory and the kanbans will be quietly shelved and replaced with the old MRP production orders. The consultants will have moved on to another client and management will be left to wonder what was wrong with the factory people who caused the lean initiative to fail.

The Nuts & Bolts of ROI and the Quarter End Shuffle

The basic arithmetic of the original ROI formula is to calculate Return On Sales and Asset Turnover, then multiply the two to arrive at Return On Investment. The example to the right shows how that works.

There have been a number of modifications to the formula over the years, but the basic principles have not changed.

Return On Investment
Basic Example

Sales	$ 1,000
Total Expenses	$ 800
Net Income	$ 200
R.O.S.	20.0%

Sales	$ 1,000
Current Assets	$ 550
Non-Current Assets	$ 600
Total Assets	$ 1,150
Capital Turnover	87.0%

R.O.S.	20.0%
X	
Capital Turnover	87.0%
R.O.I.	17.4%

Staunch advocates of ROI are quick to point out that the formula recognizes the value of lean manufacturing and is, therefore, an accurate and valid measure of a manufacturing company's performance. The assertion that the formula recognizes the value of inventory and expense reductions is true enough, but 'valid and accurate measure' is another matter entirely.

You can see that the ROI formula does show a significant benefit resulting from reducing expenses and inventory, which is the promise of lean manufacturing.

Return On Investment
Effect of Expense & Inventory Reduction

Sales	$	1,000			$	1,000
Total Expenses	$	800	$	(50)	$	750
Net Income	$	200			$	250
R.O.S.		20.0%				25.0%

Sales	$	1,000			$	1,000
Current Assets	$	550	$	(50)	$	500
Non-Current Assets	$	600			$	600
Total Assets	$	1,150			$	1,100
Capital Turnover		87.0%				90.9%

R.O.S.	20.0%	25.0%
X		
Capital Turnover	87.0%	90.9%
R.O.I.	17.4%	22.7%

The ROI formula also recognizes the benefit of expenses and inventory reductions independently, which is why the requisitions pile up on the plant manager's desk for the weeks leading up to the close of the quarter – to keep expenses at a minimum.

The next chart shows the problem that can result from increasing inventory alone – as long as there is no corresponding decrease in expenses to do so.

Return On Investment
Effect of Inventory Increase

Sales	$	1,000			$	1,000
Total Expenses	$	800			$	800
Net Income	$	200			$	200
R.O.S.		20.0%				20.0%
Sales	$	1,000			$	1,000
Current Assets	$	550	$	50	$	600
Non-Current Assets	$	600			$	600
Total Assets	$	1,150			$	1,200
Capital Turnover		87.0%				83.3%
R.O.S.		20.0%				20.0%
X						
Capital Turnover		87.0%				83.3%
R.O.I.		17.4%				16.7%

Purchased material has little or no overhead assigned to it, so there is no credit for overhead absorption when it is added to inventory. It is a pure detractor from ROI.

It is for this reason that incoming shipments are left in the parking lot, or physically unloaded, but not entered into the inventory and accounting systems on the last few days of the quarter.

It is also because of this effect that the first impressions in American of lean manufacturing – the notion that it was all about JIT, which we took to mean nothing more than shoving inventory back upstream at the suppliers – was enthusiastically embraced. For that matter, it still is in many companies.

It should be noted, however, that the significance of this in any company depends quite a bit on the company's cash situation. In a company with plenty of cash, the purchased inventory is simply swapping one Current Asset for another, so it does not make much difference to ROI. In a company that is financing inventory, it is a very big deal.

So far, so good. Donaldson Brown and his many current supporters would look at this and see no real problems. But there are problems – big ones.

Great expense reductions and opportunities to take a big bite out of inventory without impacting expenses are hard to come by. Of course, any company takes them when they can whether they are lean or not.

Unfortunately, manufacturing is rarely about the quantum leap improvements. Instead it is about the small, day-to-day activities and keeping on top of things to make sure the plant flows. The ROI formula is nothing more than algebra and there are always relationships between the variables that can be manipulated to achieve a desired outcome.

In the simple examples shown thus far – what happens to one ROI when one variable or another changes, or both change in the same direction, the ROI formula looked great. But what if we decrease expenses, and increase inventory at the same time?

As this chart shows, Donaldson Brown would think that is an excellent idea. In fact, in this basic example, ROI improves even if we put $2 into inventory for every $1 we take out of expenses.

At quarter end, plants go all out to make this happen. As noted before, simply receiving purchased inventory works the other way, but

Return On Investment
Effect of Inventory Increase & Expense Reduction

Sales	$	1,000			$	1,000
Total Expenses	$	800	$	(25)	$	775
Net Income	$	200			$	225
R.O.S.		20.0%				22.5%
Sales	$	1,000			$	1,000
Current Assets	$	550	$	50	$	600
Non-Current Assets	$	600			$	600
Total Assets	$	1,150			$	1,200
Capital Turnover		87.0%				83.3%
R.O.S.		20.0%				22.5%
X						
Capital Turnover		87.0%				83.3%
R.O.I.		17.4%				18.8%

moving purchased material along the work-in-process stream, or moving work-in-process inventory along to finished goods is quite another matter.

In those cases, adding direct labor to the value of the material means allocating overhead to the material. Any overhead that is allocated to material in production is overhead that comes out of expenses and moves to the balance sheet. Expenses go down, according to Donaldson Brown, while inventory goes up, simply by moving a skid full of material a few operations down the stream. In fact, expenses can go down and ROI can improve even when the plant pays an overtime premium to work on material that is not needed; or if the plant uses defective material in production and a large percentage of the output from production must be scrapped. For that matter, the plant can pay for all kinds of waste and see an improvement to ROI at the end of the quarter by overproducing.

Few, if any, production managers think in terms of these calculations at the end of the quarter when the all-out push is on to make sure the numbers are what corporate headquarters and Wall Street want them to be. They do not need to. The systems and subordinate measurements in the plant are all derivatives of this arithmetic. As long as labor efficiency is high, these numbers are assured. As long as the plant is performing well to Brown's flexible budgeting scheme, these numbers will come out just fine.

Perhaps the biggest problem with this scheme is that it is causing American manufacturing to get further from lean manufacturing all the time. When a manufactured product is put into finished goods inventory, the material content of that product simply moves from one inventory account to another. There is no inventory or expenses impact that arises from the material cost.

The direct labor element of the cost assigned to the product is - theoretically, at least – a wash. Assuming direct labor really is a variable cost, the cost of the labor expense is credited to inventory, so the net impact on expenses is zero.

The apparent expense reduction is all overhead. The fixed overhead, and even the non-fixed overhead that is not directly attributable to the product, is credited from expenses to inventory. In the example above, the $25 "reduction" in expenses is made up entirely of overhead that is absorbed into inventory.

It stands to reason, therefore, that the benefit to ROI resulting from building inventory is a function of the overhead content of the product. A plant with a 200 percent overhead rate will see twice the ROI improvement from building inventory as a plant with a 100 percent overhead rate. Most plants today have overhead rates of anywhere from 400 percent to 800 percent, and the number is increasing all the time. Productivity improvements through automation save direct labor cost, but increase the incentive to slow down plant flow and build inventory.

Finally, as the chart on the next page indicates, reducing inventory and allowing all of the fixed or indirect overhead that has been dumped into it to come out and hit expenses is an ROI catastrophe. For an ROI-driven manufacturer to become a lean manufacturer, sooner or later, this has to happen. Because these companies define profitability the way they do, and because Wall Street has become conditioned to believe that definition of profitability, it simply does not happen.

Return On Investment
Effect of Inventory Decrease & Expense Increase

Sales	$ 1,000		$ 1,000
Total Expenses	$ 800	$ 25	$ 825
Net Income	$ 200		$ 175
R.O.S.	20.0%		17.5%

Sales	$ 1,000		$ 1,000
Current Assets	$ 550	$ (50)	$ 500
Non-Current Assets	$ 600		$ 600
Total Assets	$ 1,150		$ 1,100
Capital Turnover	87.0%		90.9%

R.O.S.	20.0%	17.5%
X		
Capital Turnover	87.0%	90.9%
R.O.I.	17.4%	15.9%

The ROI-driven companies may make a segment of their manufacturing process look lean by driving the cycle time down and the inventory out, but the inventory is simply pushed to another part of the process. It is something akin to a python swallowing a pig. The snake can look very thin at some places, but somewhere along the length is the bloat. That is why companies like Delphi can manage to have some very lean plants, but no improvement in overall inventory turns. The inventory was simply squeezed to another part of the process. They cannot afford to get rid of it entirely. To do so would generate numbers like those on the previous chart, and Wall Street would not be happy at all.

Instead, inventory at the ROI-driven companies continually creeps up, only to come down in one of two scenarios. The first, and the one most companies hope for, is when a spike in sales volume comes along. In that case, all current production can be shipped, along with quite a bit of the excess production from previous months that has been clogging the warehouse.

The second situation is when the inventory growth has finally reached a point of absurdity. Then a layoff takes place to bring production volumes down well under demand in order to meet the demand from the excess inventory. Once inventories have been reduced, the employees are recalled, and the grand game of goosing the numbers up by sweeping overhead under the rug and into inventory begins all over again.

CHAPTER THIRTEEN

A Kanban Story

"*Bureaucracy is the death of all sound work.*"- **Albert Einstein**

Joe was the manager of a film-coating department in a magnetic tape manufacturing company. The coating department was the first step in the process, followed by a slitting operation to convert the wide, coated rolls into tape-width reels. The reels were then electronically encoded, cut to length, wound into plastic cartridges, packed, and shipped.

For a number of reasons, Joe wanted to implement a kanban process to cut the sizeable inventory between his operation and the next. For one, implementing the kanban was just good business. It would reduce inventory, save quite a bit of space and reduce the material handling and record-keeping work. For another, he wanted better visibility of what was going on downstream. He produced to a schedule, but the schedule was only as good as the forecast driving it. Sometimes the market did not behave as predicted and Joe would have to cope with sudden shifts in the mix of products in his schedule. With a kanban, he would be able to easily see the shift in demand mix and easily transition as the demand changed, rather than in an unexpected, dramatic fashion.

It took Joe six months to implement his kanban. About three days were actually required to put the kanban in place. The rest of the time went into explaining, arguing, presenting, analyzing, documenting, cajoling, and waiting. His experience demonstrates how deeply the accounting system has penetrated into the management of typical manufacturing.

For starters, the manager and employees of the winding operation viewed the proposal as one in which they had everything to lose and nothing to gain. That group was measured by whether they produced all that was on their schedules and by labor efficiency: did they get the product out within standard man hours? The mountain of inventory looked to them like security against any potential material shortage. A stock out condition meant idle labor, which would bring the wrath of upper management down on the winding people. In addition, the kanban would require additional work from them. They would have to send a signal back to Joe's people that they had used a roll of film. It took a month for them to consent, and that only came after Joe overhauled his idea a number of times to take away any effort required of the winding group and make solemn personal promises that he would look at the inventory personally every day to assure that they never ran out.

The next obstacle was the production scheduling group. First Joe had to sell the scheduler handling his department on the idea. There too, the performance measurements presented a problem. The scheduler would not get much credit for reducing inventory. He would, however, feel the weight of corporate management in the event of a failure to attain the shipping schedule. Again, there was far greater risk than benefit in the eyes of the scheduler. In addition, the work necessary to change the MRP scheduling logic to accommodate a wide open kanban was more than the scheduler was interested in doing. Another month of work went into beating down and working through all of the objections. Then the whole process had to be replayed with the scheduler responsible for the winding operations.

The next step on the tour was the plant manager, since the kanban represented a change in any number of plant policies. He required weeks of analysis, always needing more information. He was particularly concerned about 'risk avoidance' and wanted thorough research on every conceivable contingency. It should be noted that the company maintained a month of finished inventory for everything that came through the coating process. The coating machine was complicated, old and proprietary technology. A major breakdown could cause a serious disruption, hence the safety stock. The safety stock, however, meant that there was little risk of impacting customers as a result of Joe's kanban. The risk that so concerned the plant manager was the same risk that worried the winding manager. What if there is not enough material and direct labor people cannot earn standard hours for a day? That potentially could cause a noticeable budget variance. He was under no serious pressure to reduce inventories, especially not at the risk of unit cost.

The plant manager finally relented after weeks of analysis and revision. It was then up to Joe to write new work procedures for the safety, quality, and human resources people. Each of them returned the procedures a few times until he had the paperwork right.

Six months after the idea arose, Joe had his kanban.

In this brief example, everything that Sloan and Donaldson got wrong is evident.

That the most important performance measures are driven almost solely by direct labor numbers is the obvious problem. There is also a deep problem with the departmentalization. Managers have been conditioned and their jobs defined to optimize their little corner of the company without regard to anything upstream or downstream. Their objective is to make their contribution to the boss's number good – not another production manager's numbers.

Authority to make decisions is ridiculously high up the ladder, but that is where a manger has to go to find someone with the scope to change the entire process.

Staff people are all charged with controlling, not improving. Be they accountants, human resources folks, safety experts or the quality policy people, they perceive their role to be first and foremost to make sure production does not do anything wrong. They are Sloan's policy and strategy experts, while production is expected only to carry out the fruits of their thinking.

The ridiculous hoops through which people have to jump to get something done is an offshoot of Sloan style management, and was not unknown to the originators. The *Production Handbook* – a massive tome describing everything anyone would need to know about running a factory – pointed out in the 1934 edition that when working between operations, or along process lines, "contact or relationships should be initiated only with the consent of the immediate line superiors" and "before any action is taken, it must be approved by the immediate line superiors." Two production supervisors were not even supposed to talk to each other, let alone make an improvement, without checking with the boss first. Not much has changed in most companies since 1934.

This company is publicly traded, of course, and posts the closing stock price from the previous day on the plant wall for all to see (as if anyone on the production floor cared). They have made a major commitment to lean manufacturing and have spent an enormous amount of money on training and consultants. They have established non-financial measures and broadcast them widely, but they have not changed the accounting system.

There is an inventory-turns improvement goal, but it does not drill down to the individual supervisor and manager level. It is simply a broad, materials management goal. There certainly is no daily, weekly, and monthly inventory reduction tracking system with regularly planned review meetings that enable management to look over the shoulders of everyone making an inventory transaction.

The problem is that becoming lean requires a thousand projects like Joe's kanban. If each one has to justify itself against the risk of labor inefficiency, then be analyzed and discussed in detail as far up the chain as necessary to be sure that everyone whose numbers might be affected gets a chance to kill it, the company will never become lean.

The real underlying issue is that, so long as the old metrics remain driven by the cost accounting system centering on direct labor efficiency, no one believes that any new metrics or new goals supersede them. They will work towards the lean goals only after they are sure their performance to the old, core metrics is secure.

There is also a fundamental problem with the organizational structure and how authority is delegated. Because the cost accounting system measures costs from the machine level up, instead of across processes, that is how the organization is structured. Every American manager has had his or her head so packed with 'management principles' along the lines that authority and responsibility must go hand in hand that, when viewed from across the room – too far away to see the details, just the boxes – it is impossible to discern the difference between an organizational chart and a chart of accounts. In most companies, the chart of accounts simply has a name written into each summary account.

It is important to realize that at no time along the way did Joe run into accounting obstacles directly. The people in accounting would proclaim innocence in the systemic obstruction to lean manufacturing. In fact, they are probably all for it.

The obstacle was not the accounting system directly. The obstacles came from individual and departmental performance measures, the workings of the MRP system, decision-making authority, people policies, and lines of reporting that are all structured as they are because the accounting system leads management to believe that they best promote profits.

All of the management systems and structures, and the rules for running the business, have evolved in every company based on how the accounting systems indicate the company can be most profitable. The accounting system in this company defines batch manufacturing and inventory build-up as profitable. Joe did not need to talk to accounting. Every production worker, supervisor, manager, and planner he battled along the way was acting as accounting's surrogate.

CHAPTER FOURTEEN

Sloan Declares War on People

"What's the use you learning to do right when it's troublesome to do right and ain't no trouble to do wrong, and the wages is just the same?"
- *Adventures of Huckleberry Finn* - **Mark Twain**

"Engine Charley" Wilson, President of General Motors and about to become Secretary of Defense, sat before a Senate confirmation committee in 1953 and said, " What is good for the country is good for General Motors and vice versa." Like many historical facts, the circumstances and intent of this quote are often misunderstood and misrepresented. Wilson was not implying that the interests of General Motors in any way superseded those of the country. He was simply stating that their interests were one and the same. He was talking about people. The health and welfare of the people of the United States in large measure depended on the health and welfare of the great manufacturers such as General Motors.

The leaders of the big manufacturing enterprises honestly believed – and still do for that matter (with good reason) - that they were providing a real benefit to America and its citizens. They were creating jobs, making useful products, and raising the overall quality of life. What they were doing was for the benefit of many, many people, especially their

Charles Wilson did a very good job leading GM through its incredible war production effort and established himself as a very able Secretary of Defense. However, his place in history has been defined by his answer to a question during his confirmation hearing. He owned $2.5 million in GM stock that the Senate wanted him to divest in order to be confirmed. He agreed to do so, and said that he could possibly make a decision adverse to the interests of General Motors, but that he could not conceive of such a situation "because for years I thought what was good for the country was good for General Motors and vice

thousands of employees who took home paychecks and lived at a pretty good standard of life.

At the same time they were able to see the social value of business health with such clarity, they were unable to see the obvious contradiction with their methods. With a management system based around the idea that people were actually the biggest profit detractor the company had, and that, left to their own, people would gouge the company and perform as little work as possible in return, they actually saw people as the enemy of the company, instead of its reason for existing in the first place.

If Wilson's "vice versa" – that what was good for General Motors was actually good for the United States were true, it was only because General Motors had not yet figured out a way to do business without having to hire anyone at all. They tried valiantly, however, as did all of the other Sloan-type companies. The great manufacturing dream of the 60's and 70's, which is still alive and well today in many manufacturing quarters, is the idea of a "lights out" factory. To many, a factory that is fully automated and has no people, and therefore needs no lighting, is manufacturing utopia.

While deliberately breaking an integrated manufacturing process into pieces to profit from selling to inventory, instead of customers, was certainly short sighted, the rest of Sloan's scheme assured that employee involvement in production would not happen. Their approach to people, when they were given serious consideration at all, could only be described as arrogant and demeaning.

It is so commonly accepted in the United States that direct labor is a variable cost that the consequences of this arbitrary decision are rarely appreciated. At the top of the hierarchy are the policy makers, strategists and those charged with controlling the factories. Their salaries are a fixed cost, which means their jobs are relatively safe regardless of business results, within a broad reasonable range. In the middle are manufacturing management whose jobs have been, as a result of refinements to the cost accounting systems, categorized as semi-fixed or step function costs. Their jobs are only secure to a point. At the bottom are Taylor's[1] workers who should have no input into how things are made. They should be expected simply to produce to the maximum efficiency. Their jobs are purely variable, meaning their job security is purely a function of sales volume.

Those variable cost people – the ones Taichi Ohno points out are not even whole people in American cost accounting – are not people at all. They are "headcount." That simple fact makes lean manufacturing virtually impossible in Sloan companies. Lifetime employment, such as that at Toyota, is nothing more than changing the system to categorize production labor as a fixed expense.

---

[1] Frederick Taylor the father of industrial engineering said, "Management should do the planning and the thinking and workers do the work!"

A U.S. example of the same approach to people is worth considering. In the mid-1980's while General Motors was battling Toyota by acquiring robotics at a staggering rate, around the same time United Defense opened a plant in the middle of the South Dakota prairie in the town of Aberdeen. The plant manufactures products that usually get quite a bit of use when wars first begin. They make the Mk41 missile system launchers and an assortment of missile canisters for the various weapons the Navy fires from the Mk41. The destroyer based missiles that triggered the beginning of the Gulf War, Operation Enduring Freedom in Afghanistan and Operation Iraqi Freedom were all created from Aberdeen originated systems.

The early Industrial Engineers, Frederick Taylor and Lillian and Frank Gilbreth, helped to improve productivity, resulting in more and better products that enriched people's lives. Advances in manufacturing gave us better living conditions with home appliances, automobiles, healthier foods and medicines, etc. At the same time, it took the creativity from people's work. Home life was better and easier, but work life had become boring, repetitious and often hostile. We ask people to improve the quality of products and services but rarely, if ever, do we ask them to find ways to improve the quality of their work lives. The Toyota system is the light of positive change, with workers being asked for their creative ideas and being asked to participate in problem solving activities. By involving them in their jobs, we improve the quality of products and the work environment.

The plant does everything from machining, laser welding and assembly to painting and manufacturing the complex electronics units that tie the missiles and launchers to the ship fire control systems.

It is a remarkable plant that works with very little inventory, wins awards routinely for its lean manufacturing practices and quality record, routinely hits the Navy's cost reduction targets, and it is considered an extraordinary event if the plant misses a delivery commitment. It is viewed as one of the premier Navy suppliers in the country. Those things are not what make the plant remarkable. What makes it stand out is that the plant of 150 or so people stands Alfred Sloan and Donaldson Brown's theories on their ears.

The plant is entirely self-directed. There is a plant manager and everyone else is at the same level. (Technically, the HR manager holds a title for legal purposes to certify various documents required by law.) There are support people, including process and manufacturing engineers, quality specialists, materials management staff and so forth. But they have no authority over anyone else.

The plant is broken up into teams of about a dozen people. Each one has a team leader, but that person holds no authority over the other team members. The primary role of a team leader is to facilitate communications within the team and between the team and other teams in the plant.

A team also includes designated people called 'reps' that handle administrative, quality and scheduling issues. They too are primarily communicators. The teams hire, fire, give raises and mete out discipline. They are responsible for their own quality and scheduling. Everyone is on flex time, and the team members come and go as needed and decided upon by the teams. The teams measure their own performance in a wide range of areas. Teams have weekly formal meetings and numerous informal ones. Once a week there is an all plant meeting that lasts an hour or two in which all 150 people get together to discuss everything and anything of concern.

The former plant manager likened decision making in the plant to food preparation in a Chinese restaurant. For the most part, visitors are better off not looking into the kitchen to see how it actually works. His point was that it is rarely smooth, or pretty or easy to understand. There is not much hand holding and positive reinforcement going on. There is a deep culture of mutual respect that is recognized by the employees as fundamental to the success of their self-directed environment, but that does not mean they have to agree with each other, and often they do not. Meetings are typically direct, honest and blunt. Major plant-wide policy decision making is often a raucous event. There is no democracy in place. No voting takes place, except on issues such as the menu for the company picnic. Decisions are made by consensus and consensus building takes place in a wide, roundabout way.

Pay in the plant is a function of performance and skills. People earn a raise by qualifying in a technical area. Qualification means training in a skill such as machining or welding, then working in that area until the entire team agrees that the trainee is fully qualified. Beyond technical qualification, the person must demonstrate quality production, conformance with safety rules and the ability to work as a functioning member of the team in its decision making processes. It typically takes a year or two to become qualified. Once someone is deemed qualified by his or her peers, the team awards that person a pay increase. The path to another pay increase is to find a need on another team, such as welding or electronics, then make the move and start the qualification process all over again. Pay increases also require service to a team as a 'rep' in some capacity.

The most interesting result of this approach to management is that, without concern for management authority, the employee focus is radically different. A person's financial future at the company is a result of the respect, or lack of, that he or she commands from his or her peers. Whether the plant makes money or not, and whether their jobs are secure or not, is not up to a supervisor or a manager. It is up to how well the teams satisfy the U.S. Navy and other NATO customers, and the employees understand that quite clearly. They will not hesitate to reject any proposed change that they do not believe will take them closer to satisfying their customers.

Of course there are metrics everywhere, but the teams decide what to measure. They interpret them and decide whether the numbers indicate a problem and, if so, what to do about them. Engineers, supply chain people and other technical experts are involved in team issues to the extent the teams need them. If a technical support person wants to implement something new or change a process, the team must be sold on the idea, since they have all of the authority.

To satisfy the concerns of the traditionalists, the plant has no engineered standards, per se, because there is no need for them. They know how many hours per unit they averaged last year, and what goals they set for this year. Any improvements planned are based on specific, discrete projects proposed by the team, as opposed to arbitrary goals or management demands. Their measurement of themselves in terms of labor hours is just one metric of many, and not the one they worry about the most. Quality and safety goals get the most attention.

Their nearly flawless delivery record is the source of the plant's greatest pride. When a delivery is in jeopardy, having nearly the entire plant cross-trained enables the teams to lend each other as much support as it takes to be sure the delivery happens on time. As far as hours per unit are concerned, there has never been a year without an improvement, usually significant. The plant cranks out Mk41 launchers at less than half the labor hours required by the traditionally managed plant that previously built them.

The plant is far from perfect, and it is always wrestling with problems ranging from people issues and administrative rule disputes to quality problems that seem to crop up from nowhere with a product as complex as the Mk41 launcher. But they deal with the problems as effectively – usually much more so – as a top-down driven plant would, and then they move on.

The point is not that every manufacturer should run out and change over to the United Defense Aberdeen model right away – although there are worse ideas. The point is that the United Defense plant in Aberdeen proves every day that Taylor and Sloan were wrong. People are fully capable of planning their own work and setting their own policies. The plant has taken twenty years to get where it is today, and it is still evolving. It is unique in Aberdeen and United Defense has yet to duplicate it at their other facilities, although they are pursuing lean manufacturing throughout the company.

Just about every company says they want their people to work smarter not harder. Few of them understand that people cannot and will not work smarter when they have supervisors and industrial engineers hovering over them dictating and measuring their every move. They especially will not work harder if management has defined the ultimate goal to be a lights out factory, while they soar like hawks over the plant hunting for jobs to eliminate and people to lay off. People everywhere will work smarter and harder for the customer, but people will not work harder for someone who has defined them as a variable cost.

One of the keys to success at United Defense Aberdeen has been that the company has provided the plant with a good product and steady, if not increasing, demand. That is what the people ask of management. They will eagerly accept technical direction and training, and they have proven that they can continually improve the manufacturing process and compete on cost, quality and delivery with anyone. All they expect is that their efforts will not result in working themselves out of a job.

Nowhere can the inherent flaws in the traditional accounting systems and methods be more apparent than at United Defense Aberdeen. Although a few other United Defense plants are making lean progress, the company as a whole still follows the same old Sloan religion, which goes a long way to explaining why the extraordinary performance of the Aberdeen plant has not spread throughout the company. It is much the same situation as GM's inability to duplicate NUMMI. It's important to note that reports for Aberdeen stop at the plant manager's desk. Senior management has learned to leave the plant alone as long as the bottom line performance is good – and it always is.

A balance sheet prepared according to DuPont's would miss the value of the plant in Aberdeen entirely. According to the statistics in the most recent Best Manufacturing Practices award from the Navy, inventory is down 78 percent from where it was just a few years ago, and it was low by industry standards to begin with. A balance sheet, however, would not reflect that as much of an accomplishment. The same balance sheet would assign no value to the 150 cross-trained, self-directed, customer-focused employees who generate very profitable, sustained manufacturing results.

"Work smarter not harder" is a phrase often used by management and labor union leaders, but very little is really ever done to invest in workers personal growth, or to allow them to work smarter. While unions love this motto, they have insisted on retaining the job classification system, which keeps workers locked into the same job day in and day out. When I toured an Oldsmobile plant in Tarrytown, New York, I was told there was an employee who only put a tire on a hook over and over again to be lifted to the assembly line, and he had done this job for 43 years. The worst part of this story is that the worker received only two retirement checks before he passed away. Such a waste of 43 years of a human life!

Of course working smarter makes sense, but what is wrong with working harder too? Most people would rather work harder and feel some adrenaline pumping than have to stand around doing very little or the same thing all the time.

In the end, giving employees some assurance that their value is greater than short-term earnings concerns is the cornerstone of lean manufacturing. People, as the United Defense example shows, are not out to get away with taking home a day's wages for as little work as possible. The typical worker takes his career as seriously as Alfred Sloan took his. He is not about to participate in the success of a company that views him as a commodity, which variable cost designation assumes. The Alfred Sloans and Donaldson Browns of the management world view employees as something to lay off and recall at will, whenever it suits them.

Toyota revisited the original Kodak system and recognized that it was a wonderful and powerful tool to stimulate employee involvement; and that it not only saved money but made work easier, more interesting and built worker skills and capabilities as well. When I owned Productivity Press, we published a book by Yuzo Yasuda - 20 Years, 20 million ideas – The Toyota Suggestion System. In fact, last year ArvinMeritor, an automotive parts supplier in Troy, Michigan, averaged 21 ideas and $4,000 in savings per employee. A year ago, Subaru received over 108 ideas and $5,000 in savings per worker. Technicolor, in Detroit, claimed they saved over $8,000,000 from ideas submitted by their 1,800 employees last year.

Toyota believes their success rests on two pillars: lean, or the elimination of all non-value adding waste, and "Respect for People." What greater respect can you show employees than to ask them, in fact challenge them, to contribute their creative ideas; and spend a moment to listen to those ideas. And then let them implement those ideas to improve their jobs and the work environment?

Sloan made his views clear. He stated that the worker "accepts the hazards of bad times – the business cycle." He decried the "attempts to improve the economic status of our people, especially those who have so little, through hourly wage laws, shorter hours, and all the other panaceas that have been imposed on the economy" as "nothing but wishful thinking." He thought that having the workers become disciplined enough to save up for the inevitable layoff was the only solution. It was not management's problem.

"What then is General Motors?" Sloan answered his own rhetorical question, "Is it the 220,000 workers on the payrolls?" No, he answered himself, "For me, the essential ingredient – the heart if you please – of our organization is a group of not more than 10,000 workers whose skill in management, engineering and in science as well as in the special crafts make possible the work in which all others are engaged." That is Sloan in a nutshell - it was not workers who created value by making cars – it was management.

People know that any assurance of employment is only as good as the success of the business. A company cannot come through on its promise of a lifelong job if that company is failing. Most employees understand this and ask management to commit to them if things are going well.

Perhaps the fundamental lesson to be learned from Toyota and United Defense is that the Economics 101 principle of risk and return applies to hourly employees. If the company is willing to risk the possibility that there may be times when profits suffer due to an excess of people, they can anticipate an enormous return when those people commit to driving lean, high quality manufacturing through the plant. On the other hand, the company cannot have its cake and eat it too. It will not get the benefits of lean, while keeping the option of reducing headcount at will for short-term gain.

Like anything in business, lean has a risk. Toyota and the early Ford Motor Company took the risk and reaped the return. The Sloan system is built around the premise that such a risk should not be taken by modern companies.

> *How has Toyota been able to operate these past 50 years without laying off a single person?*

1. *They do have a buffer staff of part-time people. These part-time people know that there is a chance to be temporarily let go. They also know that if full-time positions open up that they will be considered first for those positions. Full-time people have a sense of security.*
2. *Toyota plans for the long term, building in all kinds of contingencies for economic business cycles.*
3. *Workers are continually trained with new skills so that they can be shifted from one part of the company to another part.*
4. *Toyota has a financial service business, information technology, other related businesses, pre-fabricated housing and operates 554 consolidated subsidiaries and 228 affiliated companies. Toyota has been continually expanding while U.S. automotive companies have been retrenching.*

> *If necessary, during slow times, Japanese automotive companies will have their workers go out and to their neighbors just knocking on doors to sell cars.*

CHAPTER FIFTEEN

People at Ford and Toyota

"It's not what you pay a man, but what he costs you that counts."
- Will Rogers

The value of engaged and supportive people is not so much to get their ideas concerning the implementation of lean practices. Their contribution certainly will be valuable, but lean does not demand their ideas to get started. People are critical to making lean work and to improving continuously the manufacturing processes once begun.

Taichi Ohno described manufacturing as something akin to the human body. Manufacturing, especially something as complex as automobile or missile system manufacturing, is an incredibly complicated network of processes and operations with millions of variables fluctuating all the time. The Toyota Production System is not a rigid, predictable scheme for operating a factory. Rather, it is a set of principles and rules that enable the factory continually to self-correct and keep on the lowest cost course when all of those variables start moving around.

Like the human body, most reactions must be reflexive and spontaneous. If the brain had to stop what it was doing to consider the best response to every change or element of danger, we would all be dead.

Ohno describes people's critical role in manufacturing as the reflexive part of the brain – the part that makes the heart start pumping harder when exercise begins to assure proper blood flow without having to stop and think about it. People, in a lean environment, will make thousands of proper decisions to cope with fluctuations in quality, machines, or human behavior. They will keep the plant on track and solve a million little problems without the need for management intervention.

That view of the role people can and must take is 180 degrees from F.W. Taylor and Donaldson Brown.[2] Another true understanding of Highland Park and how people fit in reenforces the point.

The five dollar-day-a-day program[3] at Ford, rolled out in January of 1914, is typically dismissed as a necessary move Henry Ford made to combat rampant employee turnover resulting from the dehumanizing job of working on the assembly line. Like

Toyota has rejected the concept that management does all of the thinking and the production people just work. Toyota practices "bottom-up management." The worker 'owns' 25 square feet of space and is the expert on work done in that space. Toyota believes you just have to ask the workers for their ideas, listen to them, guide them and let them implement their ideas.

In 2001, Technicolor Corporation with 1,800 employees implemented 113 employee ideas. In the last 12 months, they implemented over 16,000 ideas and saved over $8,000,000 as a result. Of course, most ideas are small and are those that the workers can implement themselves.

Imagine how workers feel about themselves and their jobs when they are allowed to think and implement their own creative ideas. This is the second pillar of the Toyota system. American management must realize the importance of involving everyone in the company in problem-solving activities to serve customers better, reduce costs, and improve quality and safety. It works, and it truly is very easy to do.

[2] They looked at people as extensions of machines just to work and not use their brains.

[3] Ford doubled the wage of workers to five dollars a day.

many of those theories discussed earlier, this simply is not the case. For one, the final assembly line did not come to life in its final form until several months after the five dollar day was announced. Even when the final assembly line was in place, a relatively small number of the total Ford employment worked on it. More importantly, the plant turnover rate had dropped to almost zero several months before the five dollar announcement.

What arrested the employee turnover rate was the implementation of a broad range of employment policies originated by John Lee, Ford's equivalent of a human resources manager and another addition to Ford management from the acquisition of the Keim Mills. The cornerstones of the policies were a formal pay-for-skills program and a high level of job security. The pay-for-skills program consisted of various job classifications depending on the skill an employee brought to the job, with laborers at the low end and draftsmen and pattern makers at the high end. Within each classification were three levels: an entry level rate, a competent rate, and an expert rate. Most striking about the policy was that the foreman, rather than the employee, was assumed to be at fault if an employee failed to progress to the highest rate fairly quickly. A foreman had to answer to top management in the rare case that an employee was not learning fast enough.

Coupled with the new pay scale was a policy that took away the absolute power held by foremen. Under the new terms, a foreman could dismiss a worker from his area, but he could not fire him from the company. If an employee was not working out, he was turned over to Lee's department which would work to find a more suitable job for that person. In the first six months the policy was in place, only one person was actually dismissed from Ford out of a workforce of over 13,000. That was as close to an employment guarantee as anyone had in the U.S. in those days. Plant absenteeism dropped from around 4 percent to under 2 percent. The five dollar day announcement came after all of this was in place and was purely a profit-sharing scheme.

To get into the plant to report for work, employees had to show a Ford identification badge. Before, during and long after the rollout of the final assembly line, described as such an oppressive work system, the employees at Ford took to wearing their company badges everywhere they went in Detroit. Whether it was to church, shopping or out socially, the people who worked at Highland Park wanted the rest of the world to know they were Ford employees.

There are conflicting stories about the spark that led to such a progressive human resources policy in a day when piece-work and Taylorism were the norm. The most reliable is that a foreman complained to John Lee about a worker who had been very good for a long time, but suddenly could not keep up. Lee talked to the man and found that he was distraught because his wife was ill, the man could not afford medical treatment, and his children were left untended while he was at work. Lee took the story to Ford and the two of them came to the conclusion that the company had a moral obligation, but an even bigger business interest, in assuring that its employees were relatively worry free. Financial worries and job security concerns turned out to be the two areas of most concern to the workers, and the new policies were aimed at resolving that.

Within a year, company schools started to teach English language and technical skills. The value of both was to enable employees to qualify for higher paying jobs.

While neither Ford, in the past, nor United Defense, today, provides a contractual guarantee of employment, the understanding with workers was and is that they would have jobs as long as the company could feasibly provide them. Any worry about unexpected layoffs is gone. At the same time, employees there and at Toyota are trained relentlessly and paid for skills, rather than production volume. As a result, even if their job does not last forever, they know they are gaining valuable skills that will enable them to work somewhere else if the worst happens. It is not difficult to gain the support of employees when they see the level of investment the company is willing to make in them.

Anyone who believes that production supervisors in any halfway decent manufacturing plant in America spend much of their time compelling employees to produce has been watching too many movies. Driving production workers – especially American production workers – to produce more is not a major part of supervision. Trying to sort through the factory chaos resulting from the myopic focus on labor efficiency is a far greater consumer of supervisory time. Dealing with machines that have been poorly maintained due to the continuing pressure to run them all the time, even just to make month-end numbers although all customer demand has been met, causes problems. Sorting through excessive work in process inventories to find the right parts can keep supervisors very busy. Dealing with quality problems also demands more shop floor attention than employee laziness does. In short, the primary role of shop floor management is not to act as modern day versions of the old plantation overseer, but to deal with the fallout from the Sloan management scheme.

People do act in their own best interests, but for the most part they are not so short-sighted as to think that their best interests are served by trying to get away with doing as little work as possible. It is in their best interests to work smarter rather than harder. This goal is often expressed by management as one the company supports, but working smarter means that someone is going to have to do some thinking. Inevitably in a Sloan-type company that task is reserved for management.

> ***Just ask yourself, "How is Toyota able to continue to grow successfully without ever having to lay off a single person?"***

If a company wants employees to work smarter rather than harder, the first step must be to kick out the followers of Fred Taylor. Their approach to industrial engineering does not require much specialized knowledge or cleverness. People are fully capable of discerning the easiest and fastest way to work. The folks actually swinging shovels under the watchful eye of Taylor already knew that shoveling some materials was harder than others. They were not so dimwitted as to need Taylor to tell than that a smaller shovel for some stuff would be better, while a bigger shovel for other material would be better.

What really took place in such cases was that management was so disconnected from the work that they needed Fred Taylor to tell them what the shovelers already knew. On the other side, the employees were certainly reluctant to suggest a way to get the shoveling done twice as fast because management's reaction would have been to thank them sincerely, then lay off half the shovelers. When a person's paycheck is a variable cost, there is no future in stuffing the suggestion box with great labor-saving ideas.

All that is really taking place at United Defense or Toyota is that the workers are empowered to pick out the best shovel for the job, instead of having to use the one management gave them. They eagerly do so because it serves to strengthen their sense of job security, rather than threaten it.

When empowered to work smarter, people will outperform the methods experts in developing the easiest and fastest way to work. Too often, however, what working smarter means in the big manufacturing companies is working according to an idea that some bright process engineer devised that he or she believes is smarter. Worker input is not being asked for; just their compliance with the new plan.

The proof of the pudding is always in the eating. Ford reduced man hours per car through a collaborative effort between workers, engineers and managers in an incredibly creative burst over a three or four year span, by as much as 80 percent. Toyota is a bit more close mouthed about such results, but Taichi Ohno let it slip once that the man hours at Toyota were more than halved during the original Corolla model run. The United Defense employees more than cut the labor cost of the Mk41 missile launcher in half.

After World War II Japan had very strong labor unions demanding pay increases from many Japanese manufacturers. At one company, the union demanded a large pay increase. The president said to the union leader, "If I grant your wish for pay increases we might be forced to go out of business. If I don't grant your wish, than we will be in disharmony. I cannot live with that so I will grant your wish and give pay increases. The union president stopped, thought and said, "We will not ask for pay increases at this time. We trust you."

When employees are a fixed cost, the source of their job security is plant profitability. When employees are a variable cost, they find job security by assuring that the work is never complete. Lean companies outperform Sloan companies because profits are in the best interests of the production employees in lean companies. It is hard to imagine how Sloan and Brown could have expected a system to work that polarized workers and management so thoroughly.

The fact that virtually every employee of every major manufacturer rushed down to the union halls when Sloan's scheme spread across the land should come as no surprise to anyone. Imagine a modern day Donaldson Brown convincing the business community today that accounting is, in fact, a variable cost; and that everyone needs one accountant per million dollars of revenue; and that there are now precise standards for cutting checks, invoicing, and making journal entries. When the sales forecast drops by $10 million, then ten accountants will be given their walking papers. The International Brotherhood of Accountants would spring to life overnight. The accounting union would walk out if the company installed spreadsheets on the finance department computers.

A culture supportive of lean manufacturing cannot emerge from a company managed under the Sloan system because the Sloan system is predicated on arrogance. The people at the top have retained the exclusive franchise for thinking and making decisions. They have further defined their jobs as responsible for controlling the rest of the people further down the chain. They have driven a management system through the company that cannot possibly support short cycle time, high quality, and low cost manufacturing performance. Finally, they have assigned responsibility for the execution of the results of their thinking to the people in the plants. Those being held responsible for manufacturing results under a set of rules that cannot possibly generate successful manufacturing have been defined as commodities - variable cost people - people who will be cast off whenever the people at the top deem them surplus to immediate needs.

In many companies the people at the top have a sense that they deserve their jobs. They earned them through years of education and hard work. At the same time, they seem to think the people at the bottom should be grateful for their jobs. This self-serving opinion is the natural result of a management system based on Sloan's values. In fact, the difference in education is taken care of by the difference in the size of the paychecks. The people at the top have not earned their jobs any more than those working in the factories. The people at the top should be no less grateful for the opportunity to earn a living than those in the factories.

No company that operates under Sloan's assumptions can attain a culture of teamwork or continuous improvement. It starts with the accounting assumption that direct labor people – head count – are variable costs. That premise drives down through performance measures, the organizational structure bent on controlling, human resource policies – even fringe benefits.

In fairness to Sloan and DuPont, they were not looking for employee participation when they devised the system. Being men of their time, they probably held an honest – terribly misguided, but honest – conviction that the men at the top *were* better men. They probably thought of themselves as not just intellectually, but morally better people than factory workers. There was a lot of that sort of judgment going around in the 1920's. They almost certainly thought of themselves as superior to African Americans; Taylor made his opinion of Italians and Hungarians clear; women were clearly not viewed as capable contributors. A compelling case can be made that the management system that Sloan devised will not work today because eighty years of human progress have proven the underlying social tenets of his system to be so wrong.

Again, his motives are up for debate, but Henry Ford exploded into the economy with a radically different view of people. Whether he honestly believed that African Americans were equally competent, or he was just trying to stick a needle in the eye of the Detroit establishment, he riled up the city fathers by putting blacks in supervisory positions over white workers. While the skill level of certain jobs often required English literacy, Ford did not deny employment to anyone on the basis of their language or nationality. He prided himself on the fact that his company was able to find or create a job for handicapped people, regardless of the nature of their handicap. He had a special program for newly released convicts. Women were included in the $5 day policy.

A cynical view would be that he did all of this to satisfy his own ego, proving his claim that he could make a contributor to society out of any man. He may have done quite a bit of it simply to irritate the elite of Detroit. Ford expressed deep hatred towards wealthy Jews. His personal bigotry might not have really been aimed at Jews as a whole, only wealthy Jews who held positions of power in banking and on Wall Street, but what came across to the public was a condemnation of all Jews.[4] He was equally brutal in his criticism of Detroit's wealthy establishment, which included most of his son's friends. He readily hired Jews to work in the factory and had no criticism of Jewish workers – just wealthy Jewish financiers. So while the rest of the country held bigoted views of one sort or another that usually centered around the lower levels of society, Ford reserved his bile for the upper levels.

Ford's motives are not as significant as his actions. When he reached his cynical and mean outlook toward life, he turned on the workers with venom, as well. But in the great days of the 1910's, people working for Ford were treated with a respect that could be found nowhere else in the industrial world.

---

[4] In 1920, Ford announced to a *New York World* reporter:

"The international financiers are behind all war. They are what is called the International Jew -- German Jews, French Jews, English Jews, American Jews. I believe that in all these countries except our own the Jewish financier is supreme... Here, the Jew is a threat."

That the Sloan system was born in a bigoted age is another, perhaps even the greatest, reason to challenge its effectiveness in the 21st century. The workers Sloan wanted headquarters to control were overwhelmingly America's second or even third class citizens. Blacks teemed from the south to work in the great industrial organizations. Immigrants poured into Detroit, Cleveland, Chicago and the other eastern and rust belt towns looking for work. Besides being culturally different, the working men were largely a hard drinking, hard swearing, fist fighting, tough bunch that gave the Alfred Sloans of society plenty of ammunition for their claim to moral superiority. If the deliberate intent of their management system was not to keep these men tightly under their thumbs, they would have been pleased to know that it had that consequence.

Modern American management holds no such racist or class biased views. At least the overwhelming majority of managers do not. Yet they are working with a system that assumes the worst of the people in the plants. At the same time, with a system under which labor efficiency creates job insecurity, they give the employees every incentive to perform at their worst.

Finally, it is very important to note that energizing people to contribute is not accomplished through warmth and fuzziness. A common trait shared by Henry Ford, Pete Martin and Charles Sorenson (the top three production people at Ford), along with Kiichiro Toyoda, Taichi Ohno and Shigeo Shingo at Toyota is that they were not particularly nice guys. They were tough, demanding and had a very low tolerance for nonsense. At the same time, they were fair (Ford was during the Highland Park heyday anyway) and treated everyone about the same. Their approach to people was no different whether they were dealing with another executive, or they with a machinist. They did not try to con anybody and they did not worry about anyone else's psyche. They judged ideas and suggestions on their own merits, without regard to the job title of the person making the suggestion.

This is not to say that a successful lean company treats people with the iron fist of Charles Sorenson or Taichi Ohno. It is saying that substituting 'feel good' management for straight talk, honest respect and fairness doesn't work.

CHAPTER SIXTEEN

R.O.S. Kills Stable Employment

"If the other fellow sells cheaper than you, it is called dumping. 'Course, if you sell cheaper than him, that's mass production." - **Will Rogers**

While many American managers would like to offer more stable, if not guaranteed, employment, it just does not seem to be possible to them. They know of no alternative other than to cut back when sales are down. The company simply cannot afford to pay for excess labor. The problem is that we are so steeped in Sloan management that we are too often blind to the fact that there are options to Sloan's principles. In fact, those options served early Ford and now serve Toyota quite well.

Another cornerstone of Sloan and Donaldson's management concept is the importance of Return On Sales (ROS). Simply put, that is the percentage of "profit", or contribution to profit, resulting from a sale. They believed they could accurately calculate the profit that resulted from selling a car at a certain price. As the theory goes, if every sale meets the ROS goal and the company makes its planned number of sales, the company will inevitably meet its profit goal, which means that DuPont would see his ROI goals met. It is yet another of Sloan's theories that works much better on paper than in practice.

There is a fundamental difference between a manufacturing company and a marketing company. Early Ford and Toyota tasked sales and marketing with selling what the factories could produce. Marketing companies – Sloan companies – task manufacturing with making what marketing can sell.

Walter Reuther led the UAW from its infancy. His growth to power, as well as the UAW's, mirrored that of Sloan and GM. In the 1950's Reuther and Sloan engaged in a running battle over the union's demand for a set annual wage – in effect, to become fixed rather than variable costs.
They also debated the impact of wage increases on GM prices. Sloan was adamant that any increase in labor cost had to be passed on to GM customers. Reuther did not agree that pricing was the only source of higher wages.
While Reuther did not look at Toyota kindly and saw them as a grave threat to American jobs, he was unknowingly advocating Toyota's labor philosophy, and wage increases through cost reduction, rather than pricing.

The damage from Brown's fixed and variable cost designation is in how it has been used to jerk manufacturing up and down. DuPont and Sloan thought the key to success was to sell a product only at the price that would provide the Return on Sales they sought. If the market was not buying, they would simply cut the factories output and wait a few months until the market was more receptive.

Ford and Toyota, once again, saw the matter a little differently. Their, manufacturing first, point of view drove them to set manufacturing in place, and then sell production at the best price they could They counted on manufacturing to be able to use workforce stability as a weapon to drive costs down. Profits would result from commanding the best price, of course, but also from continuous cost reduction. In the worst case, profits might not be as great as hoped for, but the manufacturing capability would be in place to continue the battle in the market at the lowest cost.

> *I have visited some companies that felt like armed camps. I would imagine that some company presidents do not even feel safe on the factory floor. At Toyota, Cho and other senior executives would always go to listen and learn on the factory floor. Ohno would make a manager go to the center of the plant, stand there, and watch and learn. In the U. S. senior management is often seen as the enemy and the plant floor is more like a battlefield. Just look at the recent upheaval in the airline industry.*

Ironically, driving all of that inventory into the processes, as a result of the Sloan system, pushed lead times out so far that factories have been producing to forecasts far out into the future. More often than not, the forecasts proved to be overly optimistic and there was an excess of production before the brakes could be put on. Sloan companies ended up with the worst of both worlds. The excess had to be sold at a discount, so the ROS goals were not met; and the factories were jerked down preventing the sort of manufacturing continuity needed to realize the lowest costs.

Brown and Sloan assumed that the variable costs, especially people, could be cut off with the push of a button. They believed that factories were simple machines to turn on and off, or slow down or speed up, as management desired as it chased ROI. The reality was that the factories that evolved from Sloan management were cumbersome and slow to respond.

When Sloan's followers pushed the "stop" button, the factories would slowly grind to a halt, leaving mountains of excess inventory and an increasingly disenfranchised, disgruntled work force in their wake.

Again, it is hard to comprehend the logic of the Sloan system. The bulk of the investment was in the plant and the people; yet Sloan managers somehow came to the conclusion that money could be made by shutting it all down, or at least minimizing the use of the plants and people based on market pricing conditions. They seemed to miss the obvious fact that it is impossible to get a return on a plant that is idle or people who are laid off.

> *Look at how automobiles are advertised! It often seems it is more about selling sex than well built cars. Often the ads are promising the most outlandish things, simply to get you into the showroom.*
>
> *In America, when you do decide to buy a car, you might go to the showroom planning to get one particular model, color, and option package, but soon find yourself under enormous pressure to buy whatever is in stock. In Japan, there are very few cars on the lot. You pick the model, color, style, and options and then wait just a week or so for your car to arrive.*
>
> *In 1980, I ordered a Buick and waited 13 weeks for it to arrive. A year later, on a study mission to Japan, I learned that I could have ordered a Toyota to my specifications and had it in seven days.*

Henry Ford was a neophyte concerning economics and was operating by the seat of his pants when he reduced the price of the Model T year after year. The dramatic increases in profits that resulted surprised him. He was not surprised that the volume had increased. He was only surprised by how much. He was way out in unexplored territory concerning economies of scale and the unknown size of the automobile market.

Toyota, on the other hand, knew precisely what they were doing. When American 'experts' first paid attention to them in the 1970's, they made the shocking observation that Toyota would sell at any price just to gain market share. That struck many people as just plain wrong. In any event, it was an un-American way of doing business. It is true that Toyota can be downright greedy about market share, but much of what the observers were seeing was that Toyota was going to sell its capacity at the best price above break even it could get. It certainly was not going to lay off workers or partially idle a plant because it could only sell at a 10 percent return when they really wanted a 20 percent return. Their view was, and still is, that 20 percent would be great, but 10 percent beats the pants off of 0 percent. Toyota and Henry Ford were not "buying market share." They were buying a manufacturing powerhouse. The experts steeped in Sloan's view that manufacturing is no big deal and that business is all about marketing and finance, find that impossible to understand.

The irony in Toyota's approach and the hypocrisy in the American manufacturing response was that Toyota seemed to be doing what General Motors long threatened to do in order to keep Ford and Chrysler at bay. All of the big manufacturers, not just the car makers, had institutionalized the practice of the annual price increase. There was never a question of whether prices would go up, only by how much. Ford and Chrysler understood that if either of them did not raise prices comparable to GM's increase, there would be hell to pay. General Motors stood willing to go into a price war to the finish with either of them, which could only end in ruin for Ford or Chrysler.

Then along came Toyota, seemingly willing to cut prices as often as it took to keep growing market share. This was decidedly 'uncricket' to the American companies. No one could understand such disregard for

ROS and ROI. Almost immediately came accusations that Toyota was being backed by the Japanese government in some insidious scheme to buy the American market. General Motors could match Ford, Chrysler and even Toyota dollar for dollar, but having to match the entire nation of Japan was a different matter entirely.

Toyota might have the advantage of a government relationship much cozier than that between American companies and Washington, but there was no conspiracy to ruin General Motors by devious means. Toyota was simply following the same business model as Henry Ford. Certainly they were aggressively after market share, but they were also protecting their plants first. They counted on continuous cost reductions in the plants to take care of any lost revenue resulting from price cuts. The Sloan-managed companies were simply unable to recognize a different business model. In the end, of course, market share and 'plant first' generated a much greater Return On Sales and Return On Investment than the Sloan model.

Sloan and DuPont assumed that they had enough control over the market to hold out for their asking price, which shows an amazing lack of concern for competition. It apparently never crossed their minds that the other fellow might have a different idea. They also had a false sense of power concerning their control over manufacturing costs. They designed a system based on the belief that they could turn factories on and off at will. In fact, their control systems turned the factories into slow, unresponsive behemoths.

Taichi Ohno explained very clearly that the continuous improvement of labor efficiency at Toyota resulted in cost reduction and excess capacity. The profit from lean manufacturing with a stable work force arises from freeing up people, then using them to sell more. This too is contrary to the Sloan company view. The sales forecast is based on the most the marketing department believes they can get at the planned price. Any people freed up along the way will be laid off, resulting in lower costs for the planned volume. Ohno explained that it was critical at Toyota for manufacturing to always know with precision how much extra capacity has been created, and that this information is continuously fed to marketing to update and increase the sales plan.

The distinction between a Sloan company's approach and a lean company's is dramatic. It is not possible to have the best of both. Refusing to sell unless the target price necessary to reach an ROS goal can be reached cannot co-exist with workforce stability. Workforce stability is a necessary ingredient for lean manufacturing. Again, it is the difference between a manufacturing company and a marketing company.

Few American manufacturers have made the commitment to operate with the Toyota "Production First" priority. Although that is how Henry Ford surged ahead of the pack of hundreds of smaller automobile inventor-entrepreneurs scrambling to make it in the early 1900's. Ford dominated the industry with that strategy, including General Motors, for many years until the product became hopelessly obsolete. Toyota has bested all comers with that strategy for the last thirty years or more.

The Sloan system, as applied to the relationship between sales volumes and manufacturing, has yet to demonstrate that it can compete with, let alone outperform, anyone other than another Sloan companies.

CHAPTER SEVENTEEN

Cycle Time & Cost Accounting

"Not everything that can be counted counts, and not everything that counts can be counted." **- Albert Einstein**

From a management science point of view, the practice of manipulating direct labor as the single, magic button to control cost has proven to be incorrect. To use Ohno's analogy comparing an integrated manufacturing operation to the human body, the single-minded focus on direct labor is the equivalent of selecting one vitamin or one nutrient as the magic key to health, then using it to the exclusion of all others. Vitamin C, for example, is necessary for good health and is an essential part of any diet. But a diet of Vitamin C only will kill you. Of course direct labor cost must be continually reduced, but a management system that focuses only on direct labor – or at least so disproportionately on it as the Sloan system – will ruin a factory.

Perhaps Americans are so focused on the dollars and cents of business, so intent on being 'bottom line' managers, that it is impossible for them to give much weight to any number without a "$" in front of it. However, Ford and Toyota did discover the magic bullet to control manufacturing – cycle time. This apparently non-financial number is actually the key number they used to drive just about all of their financial numbers in the right direction.

> *In Toyota's language, cycle time was the actual time it took to manufacture a product from start to finish. They came up with a new way of looking at time called takt, which is the pace of production required to meet customer demand. Then Toyota examined every operation within the company to insure that it was always running on takt. If every operation were produced at takt time, the product would always be delivered exactly on time.*
>
> *Everyone at a Toyota plant knows about takt time. When I was at Toyota Georgetown a year ago, the takt time was 55 seconds. This means that a new car must be finished every 55 seconds to meet customers' needs.*

Neither company "stumbled" across that number. It is the driver of cash flow, which was the financial priority to both companies. Taichi Ohno said in language that cannot possibly be any plainer, "All we are doing is looking at the time line, from the moment the customer gives us an order to the point when we collect the cash. And we are reducing the time line by reducing the non-value adding wastes." Ford said the same thing when he said that the only number that really mattered was the change in the bank balance from one week to the next. He was also getting at the same point when he said that "Profit is the inevitable conclusion of work well done." To the man who believed that manufacturing is a function of quality, flow and synchronization, "work well done" meant following along Ohno's time line at a pretty good clip.

Sloan and DuPont made two huge blunders. First, they unilaterally decided that they could call 'time out' whenever they chose along the time line and put material into inventory without being penalized. Second, they created a system to give them microscopic visibility to one set of activities along the line – value adding direct labor operations – at the expense of losing sight all together of the other activities along the time line.

Lean companies pay close attention to direct labor. After all, it is an important measure on the time line. They pay equal attention to everything else on the time line. Where they beat Sloan companies is that they will not save direct labor only to offset the savings elsewhere on the time line. Sloan companies do that routinely.

The power of the lean manufacturing focus on cycle time is that it simultaneously puts pressure on all of the cost drivers in the plant. To improve the time line, quality must be excellent, machines must be reliable, set up costs must be minimal. There can be no unnecessary indirect labor, and direct labor must be efficient. The key to success is to put a trained, committed work force in place along the time line, then put pressure on to urge them to continuously shorten the time line.

The weakness of the Sloan system is that, for all Donaldson Brown did to gain control, it actually creates incentives to lengthen the time line and add costs Brown could not foresee.

Cycle time is not simply a non-financial metric to look at along with labor efficiency and flexible budgets. It is the primary measure. Direct labor efficiency and direct labor expenses are a sub-set of cycle time. So are all of the rest of the manufacturing costs.

In 1980, the Yellow Book of Pennsylvania published How to Win in Manufacturing by William E. Sandman, which John P. Hayes reviewed in the Productivity newsletter in August 1980. According to Sandman, "Orders in waiting represent idle work-in-process (WIP). Reduce this waiting time and you will increase your profits and productivity."

> *Sandman's concept of productivity "is the value of the products produced relative to the investments in materials, labor and overhead required to produce those products." – ratio of output to input. He also stressed that a manufacturer's investment in materials must be used as efficiently as his investment in labor and technology. "When an order spends 95 percent of its time waiting and only 5 percent being worked on, a common occurrence, there is no balance between materials and labor." Also stated by Sandman is that manufacturing is a conversion process. It converts capital into profit, and then returns the working capital in the form of cash flow.*

Senior managers can test the cost accounting system by laying out the process for a major product, being sure that every step from arrival of raw material to shipment of finished products is on the line, then comparing it to the cost accounting information. Ask about each account – quality control, supervision, material handling, floor space related charges, inventory carrying costs, plant and machine maintenance, production scheduling – the whole gamut. Then for each account ask where on the time line it is incurred? What events on the time line cause that account to go up? Or down? What control is in place to alert management that something has happened to cause that expense to increase? The results of such an exercise should be revealing. When it comes to direct labor, the manager will be given a massive stack of reports, a list of review meetings and reams of industrial engineering analyses. For the rest of the accounts, he will get vague, theoretical responses. The accounting system most likely assumes that what drives most of the managers is direct labor, but no one in cost accounting will be able to establish that link for the manager. Others expenses will be categorized as 'fixed' or 'semi-fixed', which means that little attempt has even been made to determine what drives them. They are simply there.

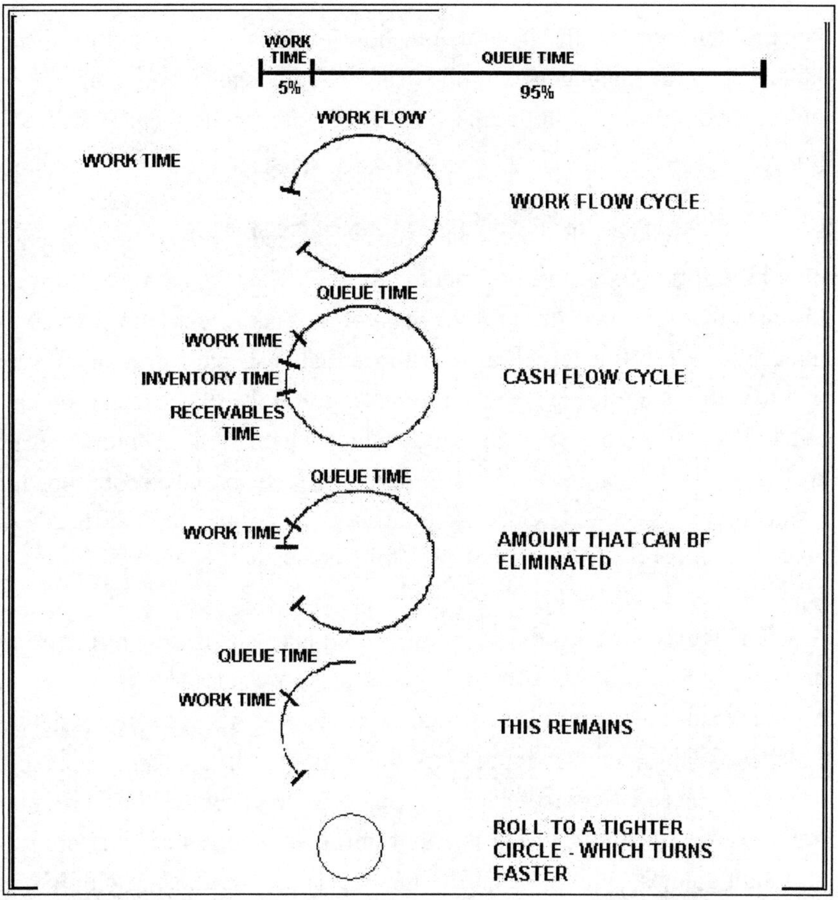

This chart was created back in 1980 and published in How to Win Productivity in Manufacturing by William. E. Sandman with John P. Hayes. It was our first awareness of the 95/5 relationship to non-value adding to value adding. It shows that paying attention to reducing the value added to non-value adding ratio is the fastest way to improving productivity and profits. It was the introduction to the concept of value stream mapping.

The lack of information about the nature of most manufacturing costs is not the biggest problem that senior managers will face. The bigger problem will be the insight into how exclusively focused on direct labor the entire management infrastructure is. Lean manufacturing requires a complete uprooting and refocus of your whole management approach.

If the company has pursued Activity Based Costing, the answers may be a little better, but not much. Activity Based Costing is built on the idea that costs are driven by things other than direct labor, which is quite true. Unfortunately, it also assumes that costs have only one driver and that the relationship between the driver and the expense can be calculated in some consistent, linear fashion, which is not remotely true. Activity Based Costing is much better than labor based accounting for product costing – it gets closer to the truth when allocating overheads to products. It is only marginally better for measuring manufacturing.

The tough news for the accounting community is that manufacturing in Detroit or elsewhere cannot be controlled by numbers reviewed in New York or anywhere else. All costs are fixed. All costs are variable. It depends on the circumstances and the timing. All costs have multiple drivers. What matters is what is driving a particular cost at a particular point in time. Manufacturing is about millions of activities and continually taking steps to assure that the unnecessary activities are eliminated and the necessary ones are executed well. The only reliable manufacturing cost number to use for measurement purposes is the total. All of the details that add up to total plant spending are interrelated and any one of them can be manipulated at the expense of the others.

Perhaps Alfred Sloan and Donaldson Brown can be forgiven in light of the fact that there was nothing like an automobile manufacturing business in history for them to use as a model. Their gross oversimplification may well have stemmed from a very real need to do something. The collection of companies and factories under the General Motors umbrella was certainly in need of order and structure. The fact remains however, that the system they came up with simply does not work.

> *Often when I speak to large audiences I am criticized for seeming to praise Toyota and Japanese management practices too much. "Why, do we have to look at Japan?" someone inevitably asks me. "Don't use Japanese words in my company," another says. "I am an American!"*
>
> *I am also an American and proud to be one. But, I want the best for us, and for the past 25 years I have been searching the world to find and share the best ideas and practices with American managers. A very wise mentor once told me, "When you want fresh milk you go to the supermarket. You look at the bottle, you want the milk to be pure, maybe you check the date, maybe even smell it. But, you don't care what the color of the cow was! You only want the milk to be fresh." Shouldn't it be the same with obtaining the best management information? You study the best in the world and you learn from them. It doesn't matter what country it comes from.*

CHAPTER EIGHTEEN

Spoon Feeding Lean to America

"It was six men of Indostan, To learning much inclined,
Who went to see the Elephant, (Though all of them were blind),
That each by observation, Might satisfy his mind.

The First approached the Elephant, And happening to fall
Against his broad and sturdy side, At once began to bawl:
"God bless me! but the Elephant, Is very like a wall!"

The Second, feeling of the tusk, cried, "Ho! What have we here,
So very round and smooth and sharp?, To me `tis mighty clear
This wonder of an Elephant, Is very like a spear!"

The Third approached the animal, And happening to take
The squirming trunk within his hands, Thus boldly up he spake:
"I see," quoth he, "the Elephant, Is very like a snake!"

The Fourth reached out an eager hand, And felt about the knee:
"What most this wondrous beast is like, Is mighty plain," quoth he;
"'Tis clear enough the Elephant , Is very like a tree!"

The Fifth, who chanced to touch the ear, Said: "E'en the blindest man
Can tell what this resembles most; Deny the fact who can,
This marvel of an Elephant, Is very like a fan!"

The Sixth no sooner had begun, About the beast to grope,
Than, seizing on the swinging tail, That fell within his scope.
"I see," quoth he, "the Elephant, Is very like a rope!"

And so these men of Indostan, Disputed loud and long,
Each in his own opinion, Exceeding stiff and strong,
Though each was partly in the right, And all were in the wrong!"[5]
- John Godfrey Saxe

Toyota magnanimously proclaimed willingness to share the secrets of their manufacturing system with competitors and the rest of the manufacturing world for the mutual benefit of all. But that is not quite how it happened. In fact, it was the stance they took after the cat was pretty well out of the bag. It seems as though they did not appreciate just how special their management and production systems were. As the trickle of curiosity from the West began to turn into a flood, they seemed to have done their best to keep the Toyota Production System under wraps until the time came when they had to concede that the factory secrets were out in the open.

When co-author Norman Bodek led his first American study missions to Japan in February 1981 to pry the secrets of their manufacturing success from them, solid information was hard to come by. Taichi Ohno allowed visitors into the Toyota plants, but cameras and note taking were

---

[5] See page 200 for the moral of the poem.

prohibited. Visitors learned more from Toyota suppliers than they did from Toyota.

> *I once asked Mr. Ohno for printed material in Japanese on The Toyota Production System that I could translate into English. He said, "We don't have anything written down for it is always changing." At the time, I accepted this - there was little else I could do - but I was sure that Ohno's reluctance was more to keep TPS a mystery to the West, for it did give Toyota a competitive advantage. Today, Toyota has a different view and is willing to share some of their secrets about their production system. Information has been available to the West for over 25 years from the books of Shingo, Ohno, Hirano and others. Toyota is more willing to teach others, especially all of their suppliers. But, Toyota knows that, in spite of all the information being made available, very few companies are willing to learn from it and apply the concepts. Most might take some aspect of the system and run a few kaizen blitzes, but have not been willing to change their attitude about people or management systems*

In the evolution of the Toyota Production System, kanban techniques and "Just In Time" deliveries from suppliers were Ohno's main contributions, although they are merely a part of the overall scheme. It was ten years or so after those efforts began that Shigeo Shingo came into the picture and kicked the whole system up to a much higher level with his work in quality control and quick changeovers. However, in the late 1970's, when Americans first began poking around trying to understand the Toyota Production System, Shingo was very much an unknown entity outside of Japan.

> *Maybe it is human nature to resist change, but so often when I came back from a study mission in Japan American managers told me that JIT only works in Japan because their culture is different. "They could get their workers to bow, to be part of work teams, to take ten minute exer-*

cise breaks, to drink green tea, to eat raw fish, to work six days a week, to average two written ideas per month to improve their work, etc." We believed that Japan was a nation of "followers," "lacking originality," while Americans are "rugged individualists," filled with "innovative ideas."

In fact, the first thing I discovered back in 1980 was Quality Control Circles (QCC), "Ah," I thought, "that was the secret of Japan's success." Truthfully, it is a very powerful process to get employees involved and bring out their innovative problem solving skills. But, QCC was only a part of Japan's secret for success, and for some strange reason most American managers still look at Quality Circles as something culturally peculiar to Japan.

How can we win the manufacturing race unless we harness the brains and the brawn of every single worker?

Before I went to Toyota on my first study mission I actually thought their factories were smaller than ours since the Japanese people are smaller. It only took one trip for me to realize that their factories are smaller because they don't need as much inventory, and their machines are smaller, too, but still produce the same product.

The population of Japan is 127 million - the 7th largest population in the world today. The size of the country is roughly that of the state of Montana with about 80 percent of the land area too mountainous to be developed. As a result, living space is extremely compressed into the 20 percent of land where all 127 million people live. Except for Hong Kong, Singapore, and Korea the population density is greater in Japan than in anywhere else in the world, yet Japan has 410,000 of the world's 720,000 "working robots."

The first cracks in Toyota's veil of secrecy came primarily from a few members of a ten person supplier team Ohno commandeered to drive the overall system through the Toyota supply chain. A plant manager by the name of Ohta at Nippondenso – Toyota's air-conditioning unit supplier – was one of the ten and he was the most forthcoming of all of the Japanese to visiting Americans. Three others from the team, Yoshiki Iwata, Chihiro Nakao and Akira Takenaka soon enlisted Bodek's assistance to start a consulting practice in the U.S. Mistakenly referred to as

Toyota industrial engineers, they actually worked for Toyota suppliers and were the sharp end of Ohno's spear in his drive to whip the supply chain into compliance with the Toyota Production System.

As Norman Bodek and the Americans accompanying him on his trips to Japan nosed around the Japanese plants in their efforts to understand the system, their first real insights came from Toyota's top tier suppliers. Ohno and his disciples certainly knew the entire system, but their forte was the kanban / JIT aspect and that is what they described to the Americans in the loudest and clearest terms. As a result of this JIT/kanban biased description, the first really authoritative book on the Toyota Production System published by Productivity in the U.S. was *Kanban – The Coming Revolution,* which came out in the early 1980's.

American manufacturing first knew the Toyota Production System as JIT – Just In Time - with kanban as the centerpiece. Taken by itself, without the context of the broader management and production system, JIT is little more than the practice of pushing inventory back on suppliers. This was the initial understanding in America, and push back on suppliers is exactly what most American manufacturers did. Lean manufacturing, of course, is much more complicated and sophisticated than JIT, however, and the results of those early efforts were insignificant.

Many U.S. manufacturers quit at that point, writing the Toyota System off as something that was overrated, or unique to the Japanese culture. A few companies still practice only this isolated lean manufacturing technique, but believe and proclaim to the world that it makes them "lean."

As America pursued JIT, Shingo began to emerge from the fog shrouding the Toyota Production System and the reality that there was quite a bit more to it than JIT began to become clear. Shingo wrote a book that went into far more detail than anything previously written, and the Japan Management Association had a limited number of copies translated into a wretched form of English – "Jangalese" - one Canadian management publication called it. Bodek cajoled the Japan Management Association into selling him a few copies of the Jangalese version of Shingo's *Study of the Toyota Production System From an Industrial Engineering Viewpoint,* as bad as it was. They demanded highway robbery prices, but sold him the books, and America's window through which the Toyota Production System could be seen opened a little wider. The complexity of the system began to come out, at least to the intrepid few who managed to struggle through the book.

It took a few years for Bodek to convince Shingo to travel to the United States, and for a clearer translation of the *Study of the Toyota Production System* to be made, but by the mid 1980's it had become plain that the *Toyota Production System* was a much greater concept than merely shoving inventory back at the supply chain. Shingo went on to write more books that were translated into English, and to travel to the U.S. extensively teaching and consulting.

Ohno ultimately wrote a book too. Actually he consented to have Bodek publish his book, The Toyota Production System, in English but it serves more to tantalize than to explain. It is filled with metaphors and theory, all very good, but its not much of a 'how to' guide, and very incomplete in its description of the overall Toyota management process.

To read and listen to Ohno and his early disciples, the Toyota Production System sounds as if it is primarily a production control system. Shingo was honest in calling his book a study from an industrial engineering viewpoint. He is perhaps the most influential innovator in manufacturing process history, as well as a great writer and teacher. Reading Shingo leaves the impression that the "System" is essentially a bag of shop floor industrial engineering techniques. Norman Bodek has traveled to Japan and visited with more Toyota people from the early years than any other American. He was left with the strong impression from his discussions and observations that the "System" revolves around human resources and organization.

Since the early revelations of the Toyota Production System, new 'cornerstones', 'paths to success' and 'critical elements' have been brought to light. "5S" and "Value Stream Mapping" are currently very popular techniques.

All of these experts are like the blind men in the parable – each partly right and all in the wrong. These unbalanced views of the Toyota Production System, each emphasized by very well qualified experts, have been amplified by the consulting, publishing and academic communities. The Internet and bookshelves are filled with 'roadmaps', 'pillars' and other mechanisms that attempt to simplify this incredibly complicated system of managing and executing a very large scale, complex manufacturing business.

Perhaps one could make the case that the "twin pillars" of an automobile are the engine and the transmission, or that the engine is the "cornerstone." They are the components that are the most complicated, most expensive and have the greatest impact on a car's performance. But so what? If the ignition switch on the car or one of the tires were left out those "twin pillars" would be nothing more than scrap metal. Lean manufacturing, like the car, consists of all of the elements. Ohno says production control is the key; Shingo says it is shop-floor techniques; Bodek says it is people. If the Toyota Chief Financial Officer had been asked, he would probably have said the cornerstone of the system was cash flow. The sales and marketing executives would have said the key was pricing. The head of product engineering would have said the entire system revolved around designing for manufacturability.

What is missing, and what none of the experts (quite a few of them self-proclaimed) seem to be able to define, is the integrated, financially driven management system that fuels all of the production control, industrial engineering and people deployment activities that are the visible elements of the Toyota Production System.

To his credit, Alfred Sloan opened the doors to Peter Drucker and allowed him free access to the entire General Motors management system. Good or bad, Drucker was free to publish it to the world. Similarly, the vast records of the early Ford Motor Company – at least those that still exist – are available to the world for the looking. Toyota still keeps its management process close to its vest. No one from Toyota has written a comprehensive study of the management of the company as Drucker did on General Motors.

The unfortunate result of this piecemeal revelation of the shop-floor workings of the Toyota Production System has been to convey the notion that it is only a shop floor concern. However, the shop-floor is not an isolated entity. It is the manifestation of management. The results of management decisions and priorities come to life in factories in the form of machines, flow and people. They are not born on their own. Taichi Ohno and Shigeo Shingo described the tip of the iceberg – the small part that can be seen. Norman Bodek and the Americans he took to Japan were also allowed to see only the tip. The tip of an iceberg is held up to the light of day by a massive, hidden base. Kanbans, Poka-yoke, U shaped cells, Multi-skilled employees, Single Minute Exchange of Dies and the other shop floor manifestations of the Toyota Production System are propped up by a Toyota Management System that is just as necessary and just as invisible as the base of the iceberg.

Both Ohno and Shingo seem to have paid a price for being as forthcoming as they were in sharing the secrets of the Toyota Production System with the world. Typical retirement for a senior manager at Toyota who has done well is a promotion to a lucrative, part-time consulting role at the executive level. Instead, Ohno was shuffled off to consult with a supplier company in his golden years. Shingo, after more than twenty years of extraordinary work at Toyota, (the value of which is probably in the billions of dollars) was given a dinner and a gold picture of Mt. Fuji, and then sent on his way. Neither was given the financial reward or the honor befitting the roles they played in creating the greatest manufacturing operation in history.

While the treatment afforded these two extraordinary men does not speak well of Toyota, the company's desire to protect the secrets to its enormous competitive advantage, is understandable. In all honesty, Henry Ford shared his methods with the world because he was an anti-capitalism capitalist. General Motors in 1960 was so big, so profitable and so powerful that they probably assumed they were invincible and that no competitor could use Drucker's information to harm them.

The elephant parable at the outset of this chapter is the most appropriate of all of the analogies used to describe American manufacturing's understanding of lean manufacturing. The mistake of each of the blind men in describing the elephant was that of over-simplification. Each thought that they could describe something as big and complicated as an elephant based on one, quickly obtained, bit of information. They each described an element of the animal accurately enough, but they all failed to comprehend the entire animal.

The lean manufacturing experience in the U.S. has been about the same. Each expert reported something different; lean was described as JIT and kanban, then as Single Minute Exchange of Dies, then as lifetime employment for involved employees, then 5S and so forth. Each element is important, but none defines the entire system.

Most important, even where the blind men get together to compare notes and assemble all of the pieces, they still would only have the external trappings of an elephant. None could conceive of the heart. The factory elements of lean manufacturing are precisely the same as the physical elements of the elephant. The heart of lean manufacturing is underneath the surface. It is lean management.

[6] See footnote for the moral of the poem on page 191.

---

[6] *Moral:*

So oft in theologic wars,
The disputants, I ween,
Rail on in utter ignorance
Of what each other mean,
And prate about an Elephant
Not one of them has seen!

CHAPTER NINETEEN

The Facts Are Clear

"We should be careful to get out of an experience only the wisdom that is in it - and stop there; lest we be like the cat that sits down on a hot stove-lid. She will never sit down on a hot stove-lid again - and that is well; but also she will never sit down on a cold one anymore."
Pudd'nhead Wilson's New Calendar
~ Mark Twain

Following their annual shareholders meeting in the Spring of 2005, under intense pressure for the continuing downward slide of Sloan's company, General Motors rolled out two initiatives to get control of the problem. The first was to lay off 30,000 workers. The second was to reduce advertised prices, but quietly, simultaneously cut out rebates and 0 percent financing. It was a sort of shell game to play with the price to lead customers into thinking they were getting lower prices when it all actually worked out to be about the same. They were doing all that a Sloan company knows to do and that is to hammer labor harder on the manufacturing side and roll out schemes more clever than the competition on the finance and marketing side.

General Electric did not mean to call horse chestnuts chestnut horses when they turned Six Sigma into the strategy to outsource and offshore everything. They were simply being a Sloan company. Six

Sigma is a business strategy, and strategy is the exclusive purview of headquarters in a Sloan company. They value-streamed the corporation and came to the conclusion that just about everyone except headquarters (of course) added insufficient value and needed to go.

Lean manufacturing has failed to take hold at every one of these companies largely because it got Sloaned. The debacle starts at headquarters with the strategy and policy makers who, after being thoroughly schooled in Toyota principles, decide that the new strategy will be that manufacturing should make their operations lean. They cannot see that Ford and Toyota ran their factories better because they managed their overall businesses in a different way. The accounting and staff people go back to their old traditional control functions and make it the responsibility of the plants to make the lean strategy happen. It would never occur to them that they, by simply doing the jobs they have been trained in business school to do, are making it impossible for the plants to become lean. It never occurs to them that the radical change needed is at headquarters first; then the plants will follow.

On the day they adopted Sloan, Brown and DuPont's management principles, all of the big companies switched from their beginnings as manufacturing companies to the finance and marketing companies they are now. It is pie in the sky thinking to believe that they can become competitive manufacturers with the same management principles and systems that drove their factories into the ground. The simple addition of a couple of new non-financial measures of performance to the same old management system is just shouting into a hurricane.

The more fundamental concern should be to question why, if the goal of manufacturing is to make money, are the things that need to be measured not already part of the financial scheme? Measurements of training and people skills, along with measures of flow are important because they drive costs and profits. They have to be added to a 'balanced scorecard' because the holes in the accounting scheme fail to consider them. The holes in the dike cannot be plugged with a couple of new metrics off to the side. The entire dike has to be rebuilt.

The thirty years from the end of World War II until the Japanese manufacturing invasion in the 1970's were a time of enormous profit for these companies, and those economic boom times are largely the reason for the standard of living we enjoy today. They came at the price of fooling ourselves into thinking we were managing well. We were not.

During those thirty years, manufacturing was in a steady state of deterioration. The lack of global competition allowed us to kid ourselves. The fact that every company was managed the same way, and followed the same ruinous manufacturing practices, kept us from giving each other the needed wake-up call. In the meantime, Sloan's terribly misguided 'concept of the corporation' was becoming more deeply institutionalized.

The practice of calling time out for inventory along the manufacturing time line in order to optimize direct labor has been validated by the SEC and, therefore, by the federal government. Brown's cost accounting ideas have become the standard for inventory valuation. Labor laws compelling separate treatment for exempt and non-exempt employees have arisen to protect those whose livelihood Brown categorized as variable costs. The curriculum of virtually every business school in the country drills "Sloanism" into the minds of the next generation of business leaders as if it were beyond question the 'one best way'. The Sloan system goes deep into the culture and laws of the United States. It will not be easy to change.

The real failure behind the inability of American manufacturing to compete with lean companies lies at the feet of the accounting community.

Isn't it time to make a major change in the whole idea of corporate identity? Isn't it time to consider that all employees, suppliers, customers and the community are stakeholders? A modern public corporation is here to serve all of those stakeholders, not just Wall Street and the stockholders. I buy and sell stock almost every week with no real thought that I am a long-term investor in the company. Is it really sacrilegious to begin to change this stockholder concept? Toyota clearly serves its management and all of its employees by investing in their growth. It serves its customers by offering great quality products, and its suppliers through loyalty and training - while also serving their stockholders. Recently my investment in Toyota climbed $40 per share making me a very happy shareholder.

The history of manufacturing accounting in the U.S. since 1923, when Donaldson Brown's concepts were adopted, is one of squabbling, infighting and turf wars with virtually nothing to show for it. The manufacturing accounting principles that guide publicly held manufacturers were almost entirely codified in ARB (Accounting Research Bulletin) 43 accepted by the SEC in 1953. The managers of the publicly held manufacturers are in a global fight for survival armed with a fifty year old accounting tool that never worked right in the first place. It's no wonder they are losing, and their efforts to function similar to Toyota has failed.

The history of manufacturing accounting since the Japanese wake up-call in the 1970's reads more like a computer technology history than anything to do with accounting. The greatest improvement accounting historians can point to since 1973 is that accounting can now be done with spreadsheets. The actual financial thinking response from accounting since American manufacturing came under siege has been silence.

Accounting has not met its responsibility to provide management with accurate information necessary to operate its factories under fire. Accounting has not met its responsibility to provide the investment community with accurate information concerning the management and health of manufacturing the companies listed on the stock exchanges.

Robert S. Kaplan and Thomas Johnson fired a clear and accurate shot at accounting with the publication of *Relevance Lost* in 1987. They made all of the weaknesses inherent in American manufacturing accounting quite clear and that was twenty years ago. Accounting has had no response. The manufacturing community, along with several academic experts, created and pushed activity based costing at the accounting community. Accounting has had no response. Now the lean community is urging managers to manage their businesses with non-financial data because the accounting data is so lacking. Accounting still has had no response.

While the manufacturers push and explore every facet of their industry looking for ways to become more competitive, ranging from robotics to self-directed work teams, accounting alone has stood unmoved in the way it has done business for over fifty years. Its (the accounting profession) collective response to Toyota has been to stand with the hand Donaldson Brown dealt them.

The Sloan advocates often cite as evidence of Henry Ford's inept business skills the fact that, every now and then, he would rampage through the accounting department, firing them all as non-contributors, and dumping their records on the floor, declaring them to be useless. To those steeped in Sloan at headquarters level, this strikes them as lunacy. To factory managers, Ford looks smarter all the time.

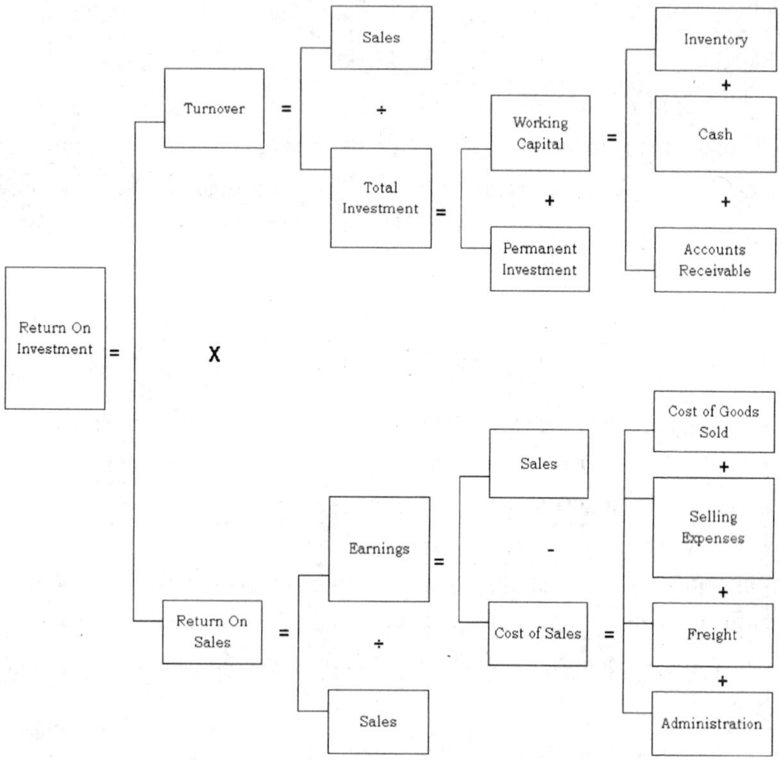

This is the logic behind DuPont's Return On Investment theory. While Toyota says all inventory is waste, Dupont says inventory is, literally, as good as money in the bank. While Taichi Ohno says that all Toyota is doing is shortening the time line between receipt of an order and customer payment, DuPont says a customer's promise to pay is also as good as money in the bank. American manufacturers cannot drive themselves to DuPont's goals and achieve Toyota's results. It is like trying to drive from Chicago to Los Angeles using a road map to New York. Without a different management model, lean efforts will continue to fail.

Direct labor is not a variable cost as a result of some mystic truth or a law of either nature or physics. It is a variable cost because management decided it would be so. Calling inventory an asset, while people are not an asset, is also a distortion of the truth. Those are simply decisions made a by a couple of men eighty years ago. In hindsight, it turns out that they were not good decisions.

The Sloan model has not served anyone well, especially the people it was intended to serve best – the investors. By focusing on ROI, the ability of companies to manufacture competitively has been undermined. There can be no significant return on an investment in a manufacturing company that cannot manufacture well. By adhering to the Sloan model, GM rode the World War II boom to the top of the industrial world, and then met competition for the first time. Its market value has gone from that greater than many small countries to a level about the same as a motorcycle company. It would be sad to think that this once great company, that has meant so much to America and the world, might ride Sloan all the way to bankruptcy before it can shake the grasp of its misguided creators.

CHAPTER TWENTY

Brass Tacks

"Remember, a dead fish can float downstream, but it takes a live one to swim upstream." - **W.C. Fields**

The facts are clear:

- No publicly held American manufacturer has become lean, except the 1-2 percent with unusual circumstances. Lean means lean at the bottom line – many large American companies have a showcase lean plant or corner of a plant for public relations purposes. Virtually every one of them has tried, and failed.

- Ford and Toyota are the two great manufacturers of the last century. No one was able to compete with Ford when Henry Ford was rational and James Couzens ran the business end of the company. No one has been able to compete with Toyota, period. Neither was a Sloan system company. Both had a cash flow focus. Both were manufacturing companies, rather than marketing companies.

- The two-edged sword of Sloan/Brown accounting – single minded focus on direct labor as the cost driver and basis for overhead allocation, coupled with establishment of inventory as an asset and a place to sell work in process – is fundamentally opposed to lean manufacturing. The two cannot coexist.

- Assigning hourly plant labor "variable cost" status is the opposite of stable employment. Viewing people as variable costs cannot coexist with lean manufacturing.

- No publicly held American manufacturer has successfully competed with a lean manufacturer, other than Harley-Davidson, Danaher, and those other few unique cases.

It is also plain that accounting does not exist in a vacuum in a manufacturing organization. The management structure of the business is a thoroughly integrated set of practices, policies and procedures all fitting together and all derived from the accounting approach.

Lean proponents agree that lean transformation requires senior management leadership and a lean culture, not just inspirational speeches and the willingness to spend money on consultants and training. The clear lesson from history is real leadership requires not just 'far behind the battle lines, speech making' type. The leadership necessary to achieve lean means having the vision and courage to dismantle the entire management system in the company and rebuild it in a manner that makes lean possible. That is a daunting task.

For many of the big, old line companies it is too late. Manufacturing is gone. For others, the task is too overwhelming. The leaders who take on this challenge are against nearly a century of entrenched thinking. They are fighting against a corporate staff whose lucrative careers have resulted from their mastery of the Sloan system. They will not give in without a fight.

The toughest battle is with finance and accounting. More than any other business discipline this group has a track record of resistance to change and creative thinking.

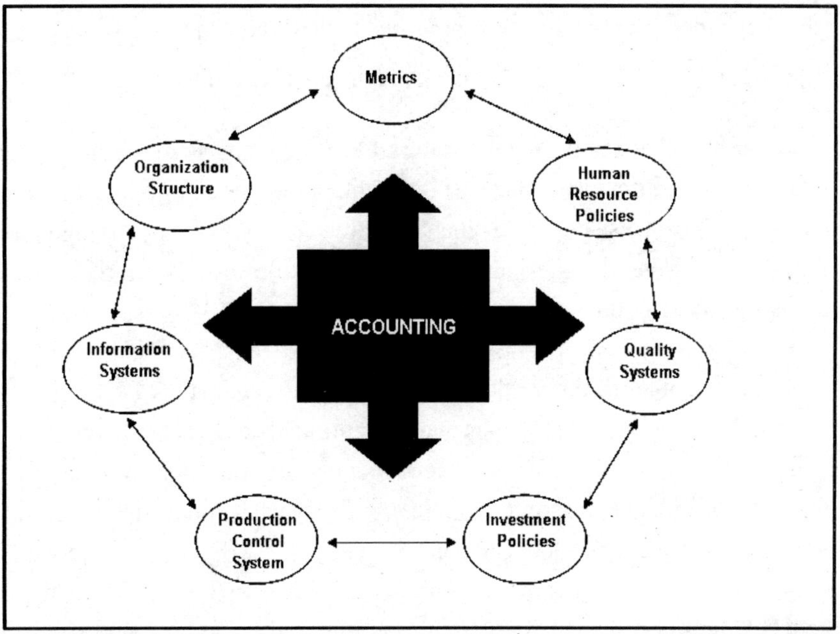

While there are many different ways to approach each element of management, the choices in any company are those that seem to best support the profit objectives of the company. Profit is defined by the accounting system. It is a matter of common sense that no company devises quality systems, production control systems, human resource policies or anything else that the financial numbers indicate are harmful to profits.

It is also true that the various elements of management are not only supportive of the accounting principles governing the company, but of each other as well. Performance metrics are structured along the lines of the organizational chart. The quality policies are built into the production control system. Policies governing what data will be collected, stored and disseminated are supportive of the organizational structure and human resources policies – some people can have access to this data, but not that data, and so forth. No well-run company allows a single department to work out of synch with the rest of the business. Every aspect of the business marches to the same music, and the drummer is accounting.

It's true that many manufacturers have their fair share of managers whose personalities drive them to be skeptical and even openly averse to change. Most resistance to change, however, is not psychologically based. It is driven by the fact that the change does not fit the rest of the management structure.

Joe's kanban initiative described previously (see pages 147 - 149) is an excellent example. He was well-intentioned enough and, by itself, kanban was a fine idea. The problem was that he was swimming against the tides of metrics, people policies, and organizational structures, as well as prevailing production control systems. All of those elements were derived from an accounting system that said batch manufacturing to achieve optimum labor efficiencies was the goal. Anyone who tried to change one aspect of the business or another into a model that supported a different objective would be in for a battle and, most often, failure.

Donaldson Brown and Alfred Sloan did not merely devise an accounting system and an organizational chart. They put an entire, integrated management structure and system in place, all geared to maximize ROI as Pierre DuPont defined it.

The history of management thinking in the United States over the last fifty years is filled with brilliant improvements when the innovation further focused and tightened this ROI driven infrastructure. MRP,

EOQ, MBO (Management By Objectives) and reams of refinements to the ROI formula itself have raised managing to a science because they all fit this model.

The history of management is also littered with seemingly great

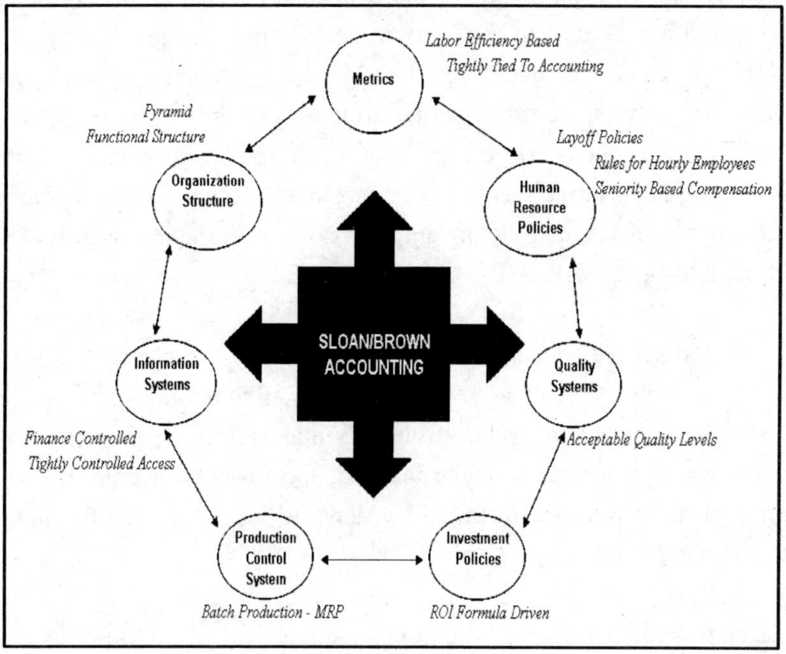

ideas that can only be termed failures. Matrix organizations, Quality Circles, participative management and JIT all failed to live up to their initial promise for the simple reason that they do not fit the current management structure. They each tried to pull a single element of managing the business out of synchronization with the rest of the system, and out of congruence with the accounting system.

People can have matrixed reporting responsibilities all over the company, but they know that the only reporting responsibility that really counts is to the box above them on the traditional organizational chart.

That person has budget authority and budget responsibility, which trumps everything else.

Access to data is always much greater at headquarters in a Sloan company because that is where the decisions are made. When the factory has limited data and the accounting system indicates that quality is insignificant until the defect rate rises to a few percent over the acceptable quality levels, then the quality circle for example is not going to accomplish very much. More important, when the hourly folks are sitting around a quality circle meeting, they are not producing. Every relevant performance metric in a Sloan company says to cut the meeting short and get them back to work.

A determined executive can force part of the management structure out of harmony by sheer force of personality if he or she wants to. That is how Six Sigma came to be. Bob Galvin at Motorola personally drove it, and was very open about the fact that the financial people had to be strong-armed from time to time in order to drive Six Sigma throughout the company.

The problem was that he did not change the rest of the Motorola management infrastructure. The day he left was the day Six Sigma began to dwindle from a way of running the factories to simply an engineering technique. It simply did not fit with the rest of Motorola's management ideal and the accounting system did not support it. It was inevitable that Motorola would revert back to its former self when he left.

The Galvin/Motorola scenario plays out time and again with almost all companies. Senior management drives an initiative that has great merit, but they do not change the basic management philosophy. After great initial results and apparent changes, either the senior managers leave or they take their eye off the ball and move on to some other initiative, and things go back to exactly where they were before.

Henry Ford, Toyota and the few truly successful lean manufacturers have a fundamentally different management scheme. Pieces of the system were assembled based on different criteria. Because the lean com-

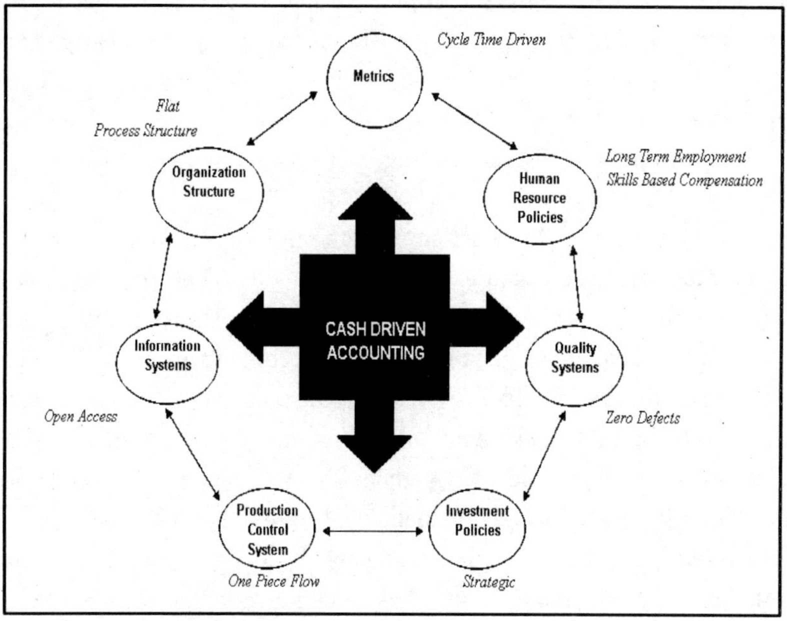

panies define profit in a different manner, the management practices are different. When Ernie Breech went to Ford to "put the GM organization garment on the Ford manufacturing frame" he converted the company from the lean infrastructure model to the Sloan model.

Had anyone asked Breech, he would have denied any intention to change how Ford ran its production and inventory control function. In all likelihood, he had no direct influence on Ford quality. Just the same, by redefining Ford finances, it very quickly followed that every element of Ford management came into alignment with the Sloan model. The old

Ford infrastructure was aimed at cash flow, while the new Ford target became ROI.

The integrated Sloan management scheme, driven by DuPont's ROI goal has become so deeply ingrained in American manufacturing management that very intelligent people repeatedly look at the old Ford records and dismiss the early Ford Motor Company as 'disorganized', 'mismanaged' and 'chaotic'. They simply cannot recognize, let alone appreciate, a management model different from Sloan's. Never mind that Henry Ford rode that 'mismanagement' to phenomenal profits.

Two great myths have arisen from the failure of management to appreciate the integrated nature of their businesses. The first myth is the incompetence of middle management. Senior executives blithely criticize middle management for the failure to become leaner, more competitive manufacturers. In fact, by leaving the existing accounting system and its congruent management infrastructure in place, senior management has directed middle management to fit a square peg into a round hole. They are asking middle management to give them Toyota factories, while retaining General Motors management. It cannot happen. In many companies, middle management has become quite jaded from having been down that road so many times. The 'flavor of the month' improvement initiative almost always means doing something in the factories that is completely against the grain of the management infrastructure. A bunker mentality sets in when middle management knows the initiative cannot succeed, and they know they will be blamed.

Middle managers cannot reduce inventories without creating an under absorption condition and resulting budget variances. They cannot make machines that were purchased purely for volume and speed become flexible and defect free. They cannot change factories when they have to stand in line behind everyone at headquarters for reports and information. They cannot cut production batch sizes overnight without incurring direct labor inefficiencies.

The second great myth is that of culture. Management is asked to make thousands of daily decisions on the factory floor: what to produce and when; whether an employee can have the afternoon off to pick up a child from school; whether to shut the machine down for maintenance or not; whether to call a marginal product defective or not; what the topic will be in the daily safety meeting.

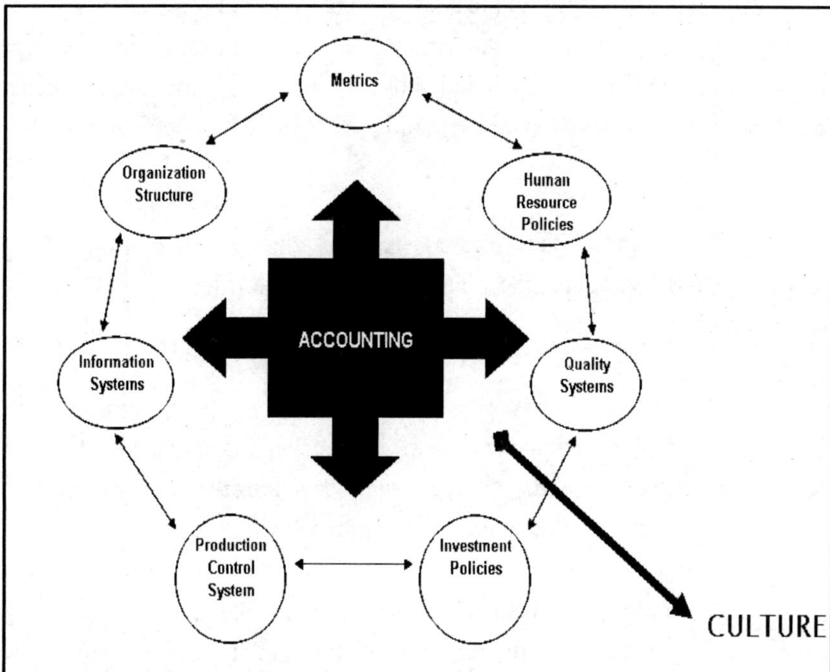

The culture of the company has almost nothing to do with the wording of a quarterly senior management speech or with posters on the wall. Mission statements are almost worthless on the factory floor. What matters are all of the thousands of personal interactions and decisions. The policies, procedures and practices of the corporate office determine how important employees really are. The relative importance of cost, quality and on time delivery are clearly communicated to people through the hourly, floor-level decisions made by team leaders and supervisors.

Factory managers and supervisors do not have the authority or the resources to run plants as they wish, especially in a Sloan company. As Messrs. Sloan and Brown wanted it, all of the resources are controlled by management decisions. People are hired or not, compensated, and disciplined when necessary according to company policies. Machines are purchased when and if approved at the top. Most supplier selection decisions are made by corporate purchasing and supply-chain folks. What to produce, when to produce it and how much to spend on which resource in order to produce are all dictated by budgets and performance measurements. Quality decisions, like people decisions, are controlled by policy.

Far from running an isolated fiefdom, the typical manufacturing manager simply executes the plan and follows the rules as laid down by management.

> *Xerox, learning a little bit about Toyota's purchasing system, decided to reduce the number of their suppliers from 5000 to 500. Did they really save money? Did they really reduce their supplier base or did they just do a little accounting shuffle?*

Lean strategies fail because manufacturing managers do not have the authority or capability to run them in a complete lean manner. Regardless of what senior managers may say, the factory will continue to execute the decisions, expressed and implied, made by the management infrastructure. In most companies, that means batch manufacturing with a dominating focus on direct-labor efficiency.

Production is carried out as a result of management decisions. The factory merely brings management to life. It is not something factory managers and supervisors do at their own whims. Teaching factory leadership the principles of lean manufacturing, then telling them to put them into practice will not generate much change. Putting those same factory leaders in a lean management infrastructure will change the factory drastically and quickly.

A lean management infrastructure will beget a lean culture and a lean factory. A Sloan infrastructure will beget a Sloan culture and a Sloan factory. It is really that simple.

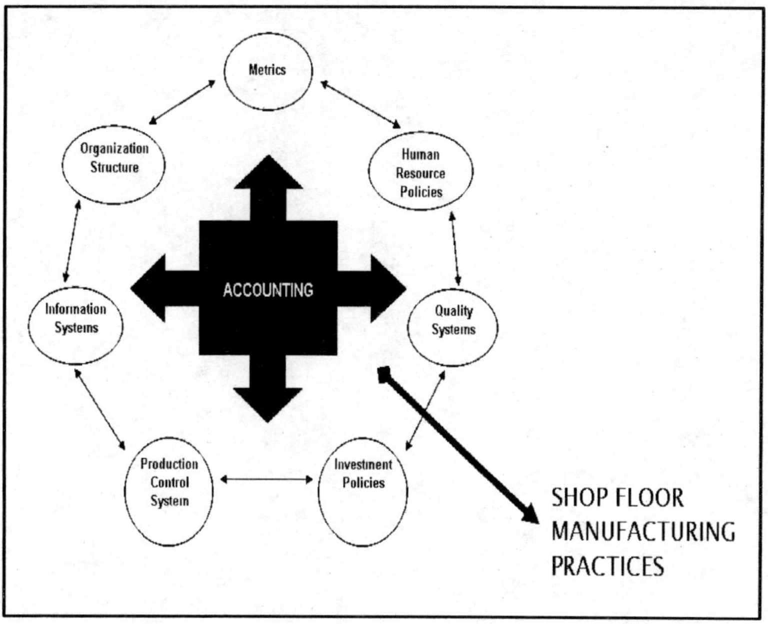

CHAPTER TWENTY ONE

Turning it Around – Step 1

"To sit back hoping that someday, someway, someone will make things right is to go on feeding the crocodile, hoping he will eat you last--but eat you he will."- **Ronald Reagan**

Drilling down to the very heart of the problem will usually get you to the wrong-headed, classical economic notion that business is basically a gargantuan struggle between capital and labor. It seems as though Pierre DuPont and Alfred Sloan missed the fundamental point that without profits, there is nothing to struggle over. They were so preoccupied with assuring that capital (i.e. the DuPont family) got to eat the whole pie and labor got none that it never occurred to them to make sure they were baking a worthwhile pie in the first place. It was the antithesis of Henry Ford's "profit is the inevitable conclusion of work well done." Sloan and DuPont leapt right to the profit end of things, apparently assuming that work would be well enough done on its own.

It should not be the primary function of management to sweep the floors every night and be sure to send every loose nickel they find to Wall Street, but that is what the Sloan system has management doing. In the end, of course, capitalism has not been served by a system designed to turn a manufacturing company into a grand ATM for investors to withdraw unlimited cash at will. Millions of investors in General Motors and the other once great manufacturers have been cleaned out. Thanks to Sloan and DuPont, their money would have been better invested in companies with funny names like "Google" and "Yahoo" than in Sloan's company.

The investors in these companies did not lose their money because they came out on the short end of a Marxist struggle with labor. The workers lost even more under the Sloan system. There are well over 1,000,000 fewer automotive manufacturing jobs in the U.S. today than there were before the Sloan companies faced Japanese competition.

Hundreds of suppliers and any number of small businesses that sprung up in the factory towns have gone under. Most companies that had a stake in the Sloan managed companies lost out. Incredibly, many of the managers of these companies still believe, for the most part, that ROI, ROS, and Donaldson Brown-style accounting are the right way to run a business.

There are clear examples of the fact that quite a good sized pie can be baked with effective management, leaving plenty for all of the stakeholders to fight over. Toyota certainly is one. In the United States, the National Football League is a great example. It is entertaining to see the millionaire players threaten to walk out on the multi-millionaire owners over the matter of just how the billion dollar profits are going to be divvied up. However, both sides seem to understand that the time for the squabble is after the season is over. Perhaps their success is based on the fact that it is not so difficult for the NFL owners and management to see the value in their workers. More likely their success stems from a common understanding that an excellent product and customer service come first, and ROI will come later. They seem to be a pretty good example of Henry Ford's statement that, "profit is the inevitable conclusion of work well done." The problem is that, in the U.S. there are few, if any examples of the same kind of success among the big manufacturers.

Wall Street should be clamoring the loudest for a different system. It is nothing short of incredible that the senior managers at Ford and General Motors can collect in January salaries, bonuses and stock options worth over $15 million for hitting the numbers Wall Street wanted, and then have their companies relegated to junk bond status four months later. Performance that brings a company to the edge of bankruptcy can only be worth $15 million in compensation if that performance was based on terribly flawed goals.

To understand the magnitude of Sloan and DuPont's debacle, the best analogy today would be to look again at one of the companies with the funny names. Imagine Google, Yahoo or Microsoft declaring that computer science is a commodity – basically any warm body from the local temp agency can do it - and that the key to success in running these technology companies is not technology, but finance and marketing. Imagine further that they all but declare war on their programmers and system design folks, classifying them as variable costs and devising a management system aimed directly at cutting their numbers and minimizing their pay.

The only way such a scenario would allow those companies to survive with such a hair-brained strategy would be (1) if all of the other American tech companies did the same thing and (2) another world war broke out and the Air Force blew up every other computer in the world. That is Sloan and DuPont in a nutshell. When the amazing events unfolded, allowing General Motors to succeed as a manufacturer that no longer cared about manufacturing, Sloan was declared to be a genius, and Donaldson Brown's concepts were chiseled into the stone of accounting principles.

As inane as that strategy is, it is still being pursued by the big companies, although they may not agree. They have in place a management infrastructure, driven by Donaldson Brown's accounting theories, that is aimed at 'controlling' the pedestrian work of manufacturing, rather than enhancing it. They cannot succeed in becoming competitive, world class manufacturing companies because they are not manufacturing companies at all.

The rebirth of American industry begins with a clear, bold decision to become manufacturers again, rather than finance and marketing firms. The factories and the people in them must be the cornerstones, with the rest of the business built around them. They can no longer be variable cost commodities to use and discard at the whims of the market.

For most companies, becoming a manufacturing focused company represents a change at headquarters. It seems to be a natural human tendency to appreciate the challenge and importance of our own job, but see the other guy's job as something a chimpanzee can be trained to do. We all think that if we were in charge of the local hamburger stand, things would be much better. We even think that if we were President, the country would be in better shape. In fact, if most of us were put in charge of the hamburger stand, the customers would all be hospitalized with ptomaine poisoning and the board of health would have us closed. Even worse, if most of us ever became President, the country would be in very, very serious trouble.

If someone knows finance very well, or holds an MBA from an excellent school, it does not necessarily follow that such a person knows the first thing about manufacturing. Going to a couple of seminars about accounting and picking up a couple of accounting books would not qualify a typical manufacturing manager to make the accounting decisions for the company. The inverse is true when it comes to the accountant who visits the plants a couple of times a year and has been to a lean manufacturing seminar.

There is a very fundamental problem in most of the big companies with the fact that the people running the business have no real idea how their products are made. Only in a business culture so deeply imbued in Sloan's concept could Roger Smith have made any sense. The man was in charge of the world's largest manufacturing company, *yet he never worked in a factory a day in his life.* The disregard for the importance and complexity of manufacturing is nothing short of incredible. While his peers in corporate management and the good folks on Wall Street may have waited with bated breath to see how his multi-billion dollar robot strategy to fix GM manufacturing would turn out, there was no suspense among America's factory managers. That colossal failure was a forgone conclusion among the people who actually work in manufacturing. The notion that General Motors could buy its way to Toyota level performance is silly. If machines were the key to success, Toyota would simply buy the same machines as General Motors. Toyota outperforms General Motors and everyone else because they out manage them.

The problem that American management must fix first is not the elimination of finance people or prevention of ideas such as GM's robotics idea. It is that the legions of manufacturing mangers slogging away in GM and the other factories every day do not have a voice. There is no one with authority at the table when such gimmicks are hatched who knows the factory environment well enough to put a stop to such nonsense.

The path to power in the golden age of Sloan management has been finance and accounting. The path to oblivion has been manufacturing. This must change immediately if any of these companies are going to have a chance to become competitive again. What passes for manufacturing experience in boardrooms these days is a year or two in a general manufacturing management position high enough up the chain that the incumbent spends most of his time in meetings and has no actual responsibility on the factory floor. What they miss are the people. A typical senior manager has no personal knowledge of what the folks working in the factory can and cannot do.

Before a senior manager can recreate his or her company into a competitive, lean manufacturer, he must remake himself into a manufacturing manager. Tom Peters made many vital points in *In Search of Excellence*, not the least of which were 'management by wandering around' and paying attention to time. Senior managers should look at their activities for the last few weeks and calculate how much of the total time they worked was spent in their factories. The typical answer is none, or perhaps a well orchestrated hour or so guided tour through one of the plants.

It should come as no surprise to anyone that a company in which the key decision makers spend five or ten percent, or even less, of their time focused on manufacturing does not manufacture very well.

What Peters called 'management by wandering around,' the Japanese call the gemba walk. It is the same thing. It means nothing more than getting out onto the factories every day for no particular reason, but for every good reason. That is the heart of the business. Hiroyuki Hirano, the Japanese industrial engineer who makes the greatest claim to being Shingo's heir apparent, talks about the "honesty of the factory." The cleanliness, the order, the flow, the look in people's eyes tells the whole truth about the manufacturing competence and performance of the business. Numbers and reports can never convey the same truth.

Senior management must start becoming a regular, unannounced presence in the plant. For the first several months, the sole purpose of that presence should be to learn by watching, asking questions and listening, but doing nothing. It is a sad commentary, but in most of the big companies it will take a couple of months of regular senior management visits to gain the trust of the people in the plants. If nothing else, the people working in the factories in a Sloan-managed company *know* that no one from senior management ever comes to the plant with good intentions. When people from headquarters make a rare appearance at the plant, layoffs follow.

> *Don't be like the General in the army who comes once or twice a year to inspect the troops. When I was in the Army we knew exactly when the General would come. That's when we started to 'polish and shine' everything. Senior leaders should be in the factories daily, if possible, to look around, to get to know their people personally, to praise their efforts, to ask questions, to listen to the answers and to encourage improvements.*

The other important activity is reading. This should not be the last manufacturing theory book read for a long while. There are shelves filled with excellent books on industrial engineering, quality control, materials management and the other nuts and bolts disciplines involved in manufacturing. The reading list should include just about everything written by Shigeo Shingo and other Toyota Production System experts.

During the routine walks through the plant, management should start to conform their questions and observations to their reading. Managers should study the lean concepts for quality control, and then learn how it is done in their factory. The same is true with pull production, versus the techniques in the plant. The goal should be to begin to get an understanding of the gaps and why they exist.

The 'why' is the toughest part. Taichi Ohno and Shigeo Shingo both advocated asking why five times in order to get near the truth. They may have underestimated the matter. If the gap between lean techniques and those in the plant seems easy to understand and easy to resolve, the real issues have probably not yet come to the surface. If it were easy, production management would have already closed the gap.

The measure of management in this first step is time. If the senior manager is still spending more time in meetings related to accounting and marketing than in manufacturing, there really is not much hope for the company. Most of the big companies have long ago cut away the fat, and their current solutions to their lack of competitiveness are cutting into the bone. A manager who has time to participate in meetings to discuss more layoffs, plant closings and outsourcing, but cannot find time to learn the manufacturing operations, is too steeped in Sloan to succeed.

The managers that do take the time to get an understanding of how manufacturing really takes place in their companies will become qualified to lead their companies through their rebirth.

CHAPTER 22

Turning it Around – Step 2

"Drive thy business or it will drive thee"
- Benjamin Franklin

Management has a choice to make. They can leave the accounting system as Donaldson Brown gave it to them or not. By leaving it alone, they can continue to be Monday morning quarterbacks – very knowledgeable and quite authoritative - but with no influence on the outcome of the game.

Sloan and Brown devised an accounting system, that was the first domino to fall in the development of the integrated management system. That management system has driven the decisions that have led to the structure, processes, practices and cultures of the factories. Manufacturing followed the system determined by Sloan and Brown.

The first option – the easy choice – is to leave the accounting system alone, which assures that everything else will remain status quo. Lean manufacturing is not possible under this option. This is the Monday morning quarterback approach. By opting to use the accounting system as it has always existed, management's input to manufacturing will largely be to look at the accounting reports and see what went wrong, instead of influencing the direction of manufacturing beforehand.

Such managers think the accounting system exists to give them numbers with which to control manufacturing. In fact, the system exists to drive manufacturing to meet certain objectives. All it gives senior management is the box score after the game is over. Middle management uses the detailed offshoots of the accounting system to make myriad, frequent, small decisions. Senior management uses it to determine who is responsible for failures.

The second option is to get in the game and restructure accounting so that it drives lean management and, inevitably, lean manufacturing. The accounting system is not a set of rules and truths divined by some higher power that must be used as management found it. Management can do just about anything it pleases with the accounting system, short of using it to steal money from the company. It is the most powerful tool management has to dictate the direction in which the company will go. As with many things in life, no decision is, in fact, a decision and usually a poor one. Doing nothing with the accounting system actually means that management has opted to leave the direction of the company alone.

A noted accounting professor once said that full overhead absorption is not a regulatory requirement. It is simply a bad habit that many accountants have picked up - similar to smoking or overindulging in ice cream or alcohol. With a little management support and a degree of self discipline, accountants can kick their bad habits, including full absorption.

Lean accounting in its purest form is accounting on a cash basis. Taichi Ohno said that all Toyota was doing was looking at the time line between receiving an order and getting paid, and eliminating the waste along the way. Lean manufacturing is all about manufacturing cycle

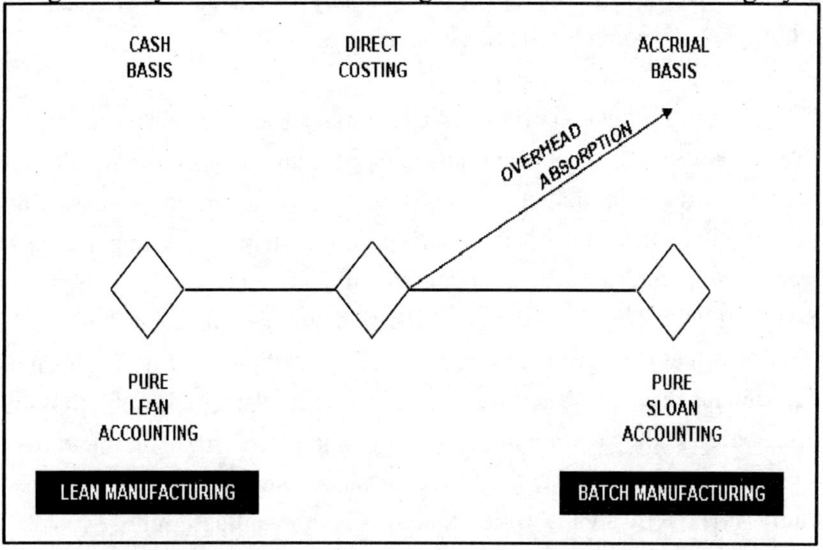

time compression and cycle time equals cash flow. Simply deciding to do all of the company's accounting on a cash basis would do the trick. The company could get rid of all of the inventory in no time, which would force production management to eliminate all of the rocks[7] of wasted expense the inventory has been hiding.

Changing the accounting procedures would be a quick, straightforward solution, but such an anti-Sloan move would, in all likelihood, be too radical and perhaps impractical for most companies. The important point, however, is that if a continuum exists between accounting that best drives effective manufacturing, and accounting that hinders it the most, pure cash flow accounting would be at one end of the continuum, while full overhead absorption accounting would be at the other end. In between lies direct costing, which is essentially taking production into inventory at its direct labor and material cost only. Between direct costing

[7] Taiichi Ohno, former VP at Toyota, showed how inventory like a river hid problems.

and full overhead absorption, the company can decide how much and what types of indirect costs to assign to inventory.

It is important to point out that Activity-Based costing and Process Costing are nowhere on the chart because they are aimed at a different matter entirely. Implementing Activity- Based Costing or Process Costing will not make much of a contribution to making a company into a lean manufacturer. Those are overhead allocation techniques to better assign indirect costs to products. Their advantage over labor-based allocations is that they are more accurate in calculating what it cost to make a particular product. They do not get at the fundamental problem, however, because they end up at the same wrong place – they facilitate overhead absorption. While they help to absorb the overhead more accurately, it is overhead absorption that kills lean manufacturing.

There is no shortage of flaws in the Sloan / Brown model, and lousy product costing is one of them. The more important objective, however, is not in knowing better what products might have contributed to the excessive manufacturing waste, but to eliminate the waste. If the factory has a poor layout, for example, requiring excessive material handling costs for anything that has to be processed on a particular set of machines, Activity-Based Costing and Process Costing will more accurately tell management which products the handling cost is wasted on. However, the waste has nothing to do with those products. Any product made on those machines will incur wasted material handling expense. The problem is the management decisions made that led to the poor layout.

In any scheme that assigns an overhead expense, such as material handling, to products by any method, the plant can appear to reduce that expense by producing quite a bit and 'absorbing' the material handling costs. Once the production is absorbed, it does not hit budgets or profit and loss statements. Instead it goes to inventory – an asset account. Because plant performance measures, and most plant decision making criteria, are based on budget and P&L impact, any accounting scheme that allows overhead to be moved from expenses to inventory creates a compelling incentive to overproduce – to build inventory. The logic behind the assignment of the overhead expense to products, whether it is labor based, Activity-Based or Process-Based, really does not make much difference in factory decision making. This is why lean accounting is cash accounting. If the plant spends money on material handling this period, it must account for it. There is no hiding it behind allowed overproduction.

Management controls the company through its decisions on which overhead expenses are allowed to be assigned to products for inventory valuation purposes. If accounting wants to assign material handling expenses to products, and credit the plants and the P&L with $X of material handling money for every item produced, senior management must ask WHY?

Production folks may not have accounting degrees, but they know a bargain when they see one. If the system says to them, you will get $5 of overhead credit for producing one more item that is not needed, but you will have to spend $1 in unnecessary quality control and material handling expense to get the $5, they will take the $5 every time. And why not? According to the accounting system, that makes money for the company. It makes no sense, but manufacturing is not to reason why – Sloan and Brown made sure that point was clear. Manufacturing is expected to execute – not think.

Every dollar of overhead that is added to the cost of a product for inventory valuation purposes increases the incentive to produce in volume, rather than eliminate waste. Henry Ford, James Couzens and the Toyoda boys played no such accounting games. Quality costs and material handling costs were going to be accounted for in full, on the spot.

The further the overhead gets from production, the more the rationalization passes from ridiculous to sublime. Donaldson Brown has everything, including the kitchen sink if the factory has one, thrown in. Plant management and administration costs, the cost of cutting the grass and plowing the snow in the factory parking lot, are all allocated through one algorithm or another. The credit to the P&L and all of its subordinate performance metrics becomes so great that the question of slowing production in order to eliminate wasted expenses is not worth asking. In fact, the deal offered by accounting to overproduce is so good that people can get fired for not taking it.

The explanation for this – the reason accounting defends the status quo – is to match expenses with sales. If production is fairly level, the profits will look a little too bad when sales are down and a little too good when sales are up. Sloshing all of this overhead expense around flattens out the profit graph.

Never mind the obvious fact – the company actually *is* more profitable when sales are strong and less so when sales are down. If senior management decisions are supposed to aim at what the stockholders would want if they were there, the call to make is also obvious. Faced with a choice between a system that can either (1) drive management and production to eliminate unnecessary costs, or (2) leave the unnecessary costs in place – in fact, encourage them to grow – but smooth the profit (or loss in too many cases) graph, the stockholders would not have to think too long before opting for choice number one.

Stockholders would opt for something closer to cash accounting for the same reason small manufacturers do. It provides better control and is much more accurate. This a 'no brainer' for small business owners because the person whose money is at stake is close enough to know the impact of this decision. This is also why smaller, privately owned businesses become lean manufacturers while big, public manufacturers fail to reach lean.

If senior management is going to lead the company to world class manufacturing levels, accounting must be compelled to come up with a much better answer than the matching principle for allocating overhead to inventory.

The second accounting rationalization is the accurate portrayal of the net worth of the company. This is as hollow as the first example. The concept of the 'value' of the company is so vaguely defined and, in most cases, irrelevant as to make just about any value placed on inventory acceptable. Is the inventory supposed to be valued at the price it might get at auction if the company goes under? Or is it the price one unit might get from a customer? Or is it somewhere in between these two extremes? Inventory is supposed to be valued at the lower of cost or market, but neither of these numbers is clearly defined. The assumptions under which the inventory might be on the market are up to the company, as are the decisions concerning which costs to include.

In reality, the value of inventory decreases as the number of units in inventory increases. Each additional unit means a little more space and a little more handling cost. Worse than that, each additional unit increases the probability that some of it is going to be excess or obsolete. It also increases the odds that the inventory accuracy is worse. At a certain point, the next item tossed on the inventory pile has no real value at all. Again, this should be looked at through the eyes of the investor whose interests should govern the decision. There is no question that the investor is best served by a conservative approach that is more apt to undervalue inventory than overvalue it, especially if that decision also supports the company's efforts to drive costs down.

Without a very good company or industry-specific justification for including some overhead expenses, the end of this exercise should be a conversion to cash flow accounting, or at least a conversion to direct costing. Allow direct labor and direct materials costs, along with other truly direct expenses such as freight, to go to inventory, perhaps, but everything else should hit the P&L and the budget in the current period.

This move alone will bring about a dramatic and instant change in factory management. The costs of quality, machine breakdowns, and material handling will come out from behind over/under absorption games and have to stand in the light.

Once direct costing is in place, manufacturing performance boils down to performance against five key metrics:

1. Total Spending (including direct labor and material) versus budget, but more important, versus the previous month.

> Management should demand continuous reduction at the bottom line level. The spending in any individual account is a plant issue and it is important that senior management leave the details alone. It is easy to create tradeoffs between spending categories. A plant can get direct labor to a very low level, so long as no one cares what happens to some of the other spending areas. The focus of senior management must be on the bottom line of total plant spending.
>
> The greatest concern people have with this approach (that everything be treated as a fixed cost regardless of production volume) is controlling costs when volume increases. They want to be sure that manufacturing spending does not increase disproportionately to volume. This is the reason for measuring to budget, as well as prior periods. Barring a radical spike in volume, management's assumption,

must be that the only increase in normal volume swings will be in direct materials. A lean plant should be able to handle normal variations in volume with the workforce in place.

The challenge for lean management will not be that the plant is spending too much on direct labor. It will very soon be that the company is unable to provide sufficient volume to the plant to handle the increased capacity resulting from productivity improvements as the plant becomes leaner.

2. Cycle Time

Cycle time must be measured from the time material arrives at the plant until it leaves the plant in the form of a finished product. It is vitally important for the success of lean manufacturing that the measurement for each product be based on the longest cycle time for any component within a product. For example, if a product requires 100 different components, the cycle time for the product is equal to the component with the longest cycle time. It cannot be a weighted average based on component cost. Using cost to weight cycle times creates an incentive to manage a few 'A' items very well, but ignore the rest. This will not lead to lean manufacturing. The only exception to this would be any 'C' items, such as fasteners, that are often purchased in bulk.

The importance of the synergy between these two metrics cannot be overstated. When manufacturing is pressured to reduce total cost and cycle time at the same time, there is no place to hide. Cost can be reduced by performing fewer machine setups, for example, but that drives up cycle time. Conversely, cycle time can be reduced by performing more frequent machine setups, but that drives up cost. The only solution for production is to reduce the setup time, which improves both cost and cycle time.

The same stress is put on all non-value adding wastes, such as material handling and quality inspection. The simultaneous pressure exerted by total cost and cycle time metrics leaves lean manufacturing as the only avenue of escape for production management.

> *On one of my last trips to Japan I met with a former author of mine Shigehiro Nakamura, who is an independent consultant and instructor at Japan Management Associations in Tokyo. He teaches and absolutely believes in "Zero Defects" – not a single defect should be passed on to the next operation. It is a very worthy goal to pursue.*

3. Customer Quality

Internal quality is a plant issue that the cost and cycle-time metrics will exert pressure on to improve. However, the plant must be measured diligently and accurately on quality from the customer. Metrics such as customer complaints, returns and warranty claims are the only effective ones.

Because internal defects inevitably result in external defects, this metric not only cuts off another avenue of escape from production management, it puts pressure on the plant to eliminate internal defects.

4. Delivery Performance

The plant must be measured on 'On Time Deliveries' with no allowances or exceptions for any reason. Note that this means performance to actual customer shipments – not to some production schedule derived from MRP or a forecast.

5. Safety

The existing plant safety metrics should be reviewed to be sure they are comprehensive and effective. Typically the standard lost time based metrics are sufficient.

A worthwhile endeavor is to put employees on safety teams and ask them find possible safety problems. Don't wait for a safety problem to occur to fix it. Don't wait for the levee to break to fix it. Go out now and look closely at all of the 'cracks' and fix them now before they become a problem.

With a direct-cost based system in place, and these five metrics (and only these five metrics – not these plus direct labor efficiency or any other relic from Sloan and Brown), the management organization must be brought around to support this new set of objectives. In most companies this will result in a painful uprooting of many deeply entrenched support groups.

Accounting, of course, will feel the greatest shock, but senior management should drive a top to bottom review of what all of the people in accounting are doing and the reports they are generating. Each job, activity and report must be reviewed with an eye toward whether it is helping manufacturing perform better to the five metrics. If not, it should be eliminated.

The same is true with the folks involved in information technology, human resources and corporate engineering. The fundamental point is that it must no longer be a corporate function to control, plan, set policy for, or strategize manufacturing. People and

JIT means delivering to the customer exactly on time - not early and not late. Toyota has been doing this for years.

departments who have been devoted to such tasks should either be sent to the plants or eliminated.

Ford recently announced that they are eliminating 4,000 such positions. The people in those jobs, as in many corporate functions under the Sloan model, very often are the best and the brightest. The 4,000 folks who are falling under Ford's axe are probably very capable, intelligent, educated and hard working people. They are not losing their jobs because of their own doing, or because of a failure in Ford marketing or finance, or because of the UAW or the American regulatory environment. They are losing their jobs because Ford should have had their best and brightest focused on manufacturing excellence. Instead, they were focusing their best resources elsewhere.

The net effect of all of this should be a redirection of the resources used to exert Sloan and Brown's centralized control to the plants, along with a set of ironclad, bottom line metrics that demand excellent manufacturing performance.

The final move in the rebirth of American manufacturing is to put the company's money where its mouth is when it says 'people are our most important asset'. It is nothing short of hypocrisy for a company to carry destructive, over-valued inventories on the balance sheet as assets, while people are reflected nowhere other than on the expense column of the profit and loss statement. Any accounting argument that carrying inventory at full absorption rates so an investor can know the honest value of the company, while ignoring people, rings very hollow.

Sloan and Brown were dead wrong. People are not interchangeable commodities, to be hired and laid off every time the wind blows from a different direction. Manufacturing is too complicated; there are far too many variables. There is no computer big enough or fast enough to plan and control it. Toyota knows this. It takes a factory full of trained, focused, committed people to get all of the details right in the midst of so many dynamic events. They are not commodities.

A new accountant can be hired and pretty much become as effective as he or she is going to get within a month or two. A production employee is more likely to take six months or more – a year according to the self directed teams at United Defense. Yet the accountant is a fixed expense with a fair amount of job security, while the production worker is a commodity.

As discussed previously, when the plant is sent the message through the accounting system that activities that move money from expenses to the balance sheet are profitable, they will push those activities very hard. The best way to develop the work force quickly and increase balance sheet accuracy for the sake of the investor is to capitalize the cost of training and educating people.

Whether the company wants to capture the cost in detail, or simply take a percentage of every employees wages and move it to the balance sheet is not as important as creating a mechanism that reflects the value of people.

When Ford loses those 4,000 people, the company will be worth a little less than it was with them. They reflect an investment in training and experience that will cost the company quite a bit to replace. While they were working, the value of such resources should have been reflected in the statement of the value of the company. When they are let go – when Ford cuts a little deeper into the marrow of the company – they should take a write off so the investors know the impact.

Only in the world of F.W. Taylor, Pierre DuPont, Alfred Sloan and Donaldson Brown can kicking trained, experienced, capable people out of a company be seen as a positive move. There is nothing positive about it. It is proof of a basic failure of management. To do so within months of paying the top manager better than $10 million in performance bonuses ought to be proof enough that the system is broken.

There is often a fair amount of skepticism on Wall Street's part when a company capitalizes something intangible, and making the bold move to put people on the balance sheet will certainly raise some concerns. That has never stopped the Sloan companies from capitalizing goodwill and intellectual property, often on far shakier ground than capitalizing people. It is entirely within the purview of the company, however, and the choice is management's. All that is required is a reasonable, consistent, logical method of doing so.

CHAPTER 23

The Call to Action

"There is nothing wrong with America that cannot be cured by what is right with America"
- President Bill Clinton

This treatise has been harsh on the great men who created the most powerful economic engine in history. It has also been harsh on some of the men and women who have earned their way into leadership roles in some of America's greatest institutions. At the same time, the deterioration of great segments of America's manufacturing economy has been much harsher on millions of people. There is a fundamental problem that cannot be solved by tweaking the old system.

There is an entirely new concept of manufacturing that has proven itself at Toyota. It is spreading throughout the world, but America seems to be having a very hard time 'getting it'. If we do not 'get it' soon, the economic consequences to future generations of Americans will be even harsher yet.

The course of action outlined for management is a very difficult one, but entirely doable. Unlike the challenge faced by Henry Ford and Eiji Toyoda, there is a roadmap available. This is simply the first call to action. Others, wiser than the authors, will certainly take these ideas and build on them, making improvements where we left holes. In the end, a new management model that will do where DuPont and Sloan didn't succeed will emerge. The next Eliji Toyoda is certainly waiting in the wings of one of these great companies to take the lean management ball and run with it.

There are a few basic principles management must take from this book. First, accounting is a management tool to be used as the primary vehicle to direct the company in any direction leadership wants to take it. Left unchanged, however, the accounting system will continue to drive the company to pursue the same manufacturing practices that have failed in the past. With the same accounting and management systems, about the only thing senior management can do to prevent ineffective batch manufacturing in their plants is outsource.

Second, accounting drives the entire management infrastructure. It will form a cohesive, congruent set of disciplines driving every aspect of the company to meet whatever objectives the accounting system places value on. No alternate set of metrics or factory level lean manufacturing initiatives are going to knock the company very far off the course the management infrastructure is designed to take it.

Next, Ford and Toyota were driven to develop and aggressively pursue lean manufacturing because they had fundamentally different management infrastructures that naturally flowed from the fact that they were cash driven, rather than ROI driven.

Finally, in order to become a lean manufacturer, a company must first become a lean managed company. This is no small task. It requires radical change to the accounting system, adopting an entirely new set of measurements from the accounting system, shattering and recreating the entire management structure, and making a commitment to the people of the company. That will be a daunting endeavor for even the toughest and most visionary of managers.

Each person must be given the opportunity to grow on the job, to go beyond the "dead end" jobs, to improve their skills. As the bible says, "We are created in the image of the divine." Each of us should be given the opportunity to express our creativity at work. If Toyota and other Japanese companies can do it – so can we."

While only the top management of the big American manufacturing companies can make lean manufacturing happen in their organizations, there are other entities within the organization that have an obligation to contribute.

An even more convincing book can be written demonstrating to organized labor that the model they have used has not worked for employees any better than the Sloan model has worked for management. While many of the positions taken by labor unions have been forced upon them by Sloan management, when the days of Sloan end the unions must recognize this and adapt. Manufacturers must make a strong commitment to cross training and deeply training their employees. This means the end of rigid job classifications. People should be able to perform a broad range of jobs across the factory spectrum. People will also be paid for their skills, not their seniority. Those two basic principles of lean manufacturing must become realities in America's manufacturers, or the companies will not be able to compete and jobs will be lost. Labor has an obligation to rethink its values and contribute to the model that enables employers to become lean manufacturers.

If they do not, organized labor will simply cease to exist in the manufacturing sector. Either manufacturers will become lean by kicking the unions out in order to work with their employees effectively; or the unions will prevail and jobs will all be outsourced. Either way, the labor organization that does not go through its own rebirth will not long survive.

Government also must decide to support manufacturing. While the accounting and management principles outlined in this book are legal, they will require companies to swim upstream against the current of IRS and SEC thinking. The regulatory agencies must develop policies that make lean management and more cash based accounting the norm; and expect to see people on the balance sheets.

Most important, elected officials must learn how manufacturing actually impacts the economy, then become leaders and statesmen, rather than politicians. The capital markets are certainly very important to the economic health of the nation, but the nation's leaders seem to have a grossly distorted view of the significance of the Dow Jones index. It is not a crisis when the New York Stock Exchange falls by a few points – even by many points. It is an economic crisis when one manufacturer after another shuts the doors and moves the work to India. It is nothing short of an abomination when political leaders open their office doors to CEOs who are openly bailing out of America. It is equally ridiculous when other leaders kowtow to union leaders locked into practices that suppress their workers and prevent companies from succeeding.

Policies and laws should be based on the input of manufacturing people committed to the rebirth of industry in America, rather than any narrow interest.

The investment community wields more power than anyone in driving a regeneration of manufacturing. The investment community should be more concerned than anyone with the fact that they are putting money into companies that continually lose market value, despite well thought out strategies and seemingly healthy numbers. It was the investment community that paid the top people at Ford and GM such hefty sums, then gave their companies 'junk bond' classification a few months later. There should be total agreement with the notion that, despite best intentions, Sloan and DuPont got it wrong.

Finally, the academic community is perhaps in the best position to lead the rebirth of American manufacturing. Unencumbered by personal agendas and not beholden to Wall Street, the leading business schools are in an excellent position to objectively look at Sloan, and explore the facets of the accounting and management systems of the 21st century. They also should be concerned that their 'best and brightest' graduates have been unable to influence the direction of the once great manufacturing companies.

There is no question that America will see through the smoke and confusion of the changing economic world and lead manufacturing again. There is too much human energy and capability in this country for it to be otherwise. The rebirth of American industry must come and it will come, after all, manufacturing is vitally important. We thank you for reading this book and we challenge you to take leadership in our rebirth.

* * * * * *

I see America, not in the setting sun of a black night of despair ahead of us, I see America in the crimson light of a rising sun fresh from the burning, creative hand of God. I see great days ahead, great days possible to men and women of will and vision.- **Carl Sandburg**

INDEX

References:

Henry Ford is perhaps the most widely written about American after Abraham Lincoln. There are literally thousands of books and articles written about him over the last 100 years. Most are for the popular reading market and limit their discussion of Ford manufacturing to the assembly line.

Ford is loved or hated with a passion even though he has been dead for over fifty years. Many authors seem to have a strong emotional bias – mostly negative – that should be taken into account by the reader. In large measure, he brought the ill will upon himself with his anti-Semitic rhetoric.

The most reliable sources of information about Ford and the details of early events at the Ford Motor Company are three volumes by Allan Nevins and Frank Hill, entitled *Ford: The Times, the Man, the Company* (1954), *Ford: Expansion and Challenge*, 1915-1933 (1957), and *Ford: Decline and Rebirth,* 1933-1962. The first volume has been particularly helpful in writing this book.

A tedious but reliable and thorough reference to the management and manufacturing operations of Ford in the early days is Arnold and Faurote's *Ford Methods and the Ford Shops* (1915). They had almost unlimited access to the plant at Highland Park during the first year after the last assembly line was installed.

A book that is both reliable and interesting from the standpoint of Ford's broader life is Robert Lacey's 1986 book, *Ford: The Men and the Machine.*

Finally, there are the books Ford wrote himself, including *My Life and Work* (1922) and *Today and Tomorrow* (1926). Both were ghost written by Samuel Crowther, but they seem to reflect the true, if somewhat more eloquent, sense of Ford's thoughts. *Today and Tomorrow* is particularly significant because it is the one most often cited by the early Toyota management as having had a big influence on their thinking.

Other Ford references, including Harry Bennett's *We Never Called Him Henry* and Charles Sorenson's *My Forty Years With Ford* should be read with a bit of skepticism. Both books seem to be more of an attempt by the authors to salvage their personal reputations than to accurately document events. Bennett's reputation needed all the help he could give it, but it is disappointing that Sorenson took this approach. He led an incredible manufacturing life and could have offered much to the body of knowledge.

Alfred Sloan wrote his work autobiography in 1963 entitled *My Years With General Motors*. Another book that provides quite a bit of insight into Sloan's thinking is his *Adventures of a White Collar Man*, written in 1941. This book is especially interesting if you also read the essays written by Walter Reuther in the 1940's and 1950's. Sloan's book and Reuther's essays constitute an interesting debate. There is also quite a bit of information available in the General Motors corporate history files. For the most part, however, in addition to creating GM's management system, Sloan set the standard for the staid, low profile, personally dull manager that has come to symbolize General Motor's Sloan's culture. The academic community has written extensively about his management system, but the popular press largely ignored him. He provided them with none of the color of Henry Ford.

An Internet search of any aspect of Sloan's business model will bury you with more academic research papers than you can absorb. The "M" form corporation and Sloan's concept of centralized control and decentralized responsibility is perhaps the most widely studied and analyzed concept in business history. Explore as you think best and form your own opinions concerning which documents offer valuable insight. The conclusions drawn and opinions expressed in this book are entirely my own and do not reflect any other person's research into Alfred Sloan.

Donaldson Brown was much like Sloan in that his personal life offered nothing of interest to the press. Much of his work actually began at DuPont and he brought several well developed ideas to GM. Both DuPont and GM company histories include considerable information about Brown, his contribution to accounting and the details of his career. He wrote two significant white papers, which have been cited and reprinted in numerous college text books and accounting journals. *Pricing Policy in Relation to Financial Control* was written in 1924. *Decentralized Operations and Responsibilities with Coordinated Control* followed in 1927. Both are tedious documents that only an accountant can love. They also lend considerable credibility to the assertion by some of Brown's peers that he was so intelligent that they often could not understand him. The documents are in the General Motors corporate history, but are also generally available in whole or in part through colleges and accounting organizations.

Robert Kanigel wrote a wonderful biography of **Frederick W. Taylor**. It is a great credit to his writing skills that he makes a lengthy book about scientific management interesting. *The One Best Way: Frederick Winslow Taylor and the Enigma of Efficiency* was written in 1997 and is a great manufacturing book for anyone. If that does not satisfy the appetite for Taylorism, Taylor's own writing is available. *Shop Management* was written in 1903 – the same year the Ford Motor Company was created. *Scientific Management* came out in 1911. Neither is in print to my knowledge.

The best one volume book on **Frank Gilbreth** is *Frank and Lillian Gilbreth, Partners For Life* written by Edna Yost in 1949. It is a little too heavy on the praise, but a generally factual recitation of their careers. *Applied Motion Study* By Frank & Lillian Gilbreth written by the couple in 1917 provides a good understanding of their thinking.

Shigeo Shingo wrote a number of books that have been translated and published in the United States by Productivity Press. *A Study of The Toyota Production System from an Engineering Viewpoint, Non-Stock Production,* and *The Shingo Production Management System* should be mandatory reading for anyone attempting to get a genuine understanding of the Toyota Production System. A brief but interesting biography of Shingo has been published by the IES program at North Carolina State University.

Taichi Ohno's *Toyota Production System* should also be at the top of the list for anyone pursuing lean manufacturing. It is a book to be read several times. It appears to be a simple, quick read, but almost every word had deep significance.

Japan Review published a fascinating article entitled *The Introduction of Scientific Management* to Japan in 2002, by Hashimoto Takehiko. In addition to industrial engineering, the article provides a rare look at the relationship of work to life in Japan, and the relationship between employee and employer.

The New York State Society of CPAs published a two part history of accounting in the CPA Journal in January and February 2005. In addition, Texas A&M University has compiled a considerable history of accounting. These two sources should provide ample reference material for all but the most passionate accountant.

William H. Waddell
www.bestmanufacturingpractices.com

Email: bill@bestmanufacturingpractices.com

Bill Waddell is an independent lean manufacturing consultant, author, editor and teacher. He has been a lean manufacturing practitioner for over 20 years. His published works include numerous articles on lean manufacturing. His regular lean blog on Superfactory regularly draws over 1,000 readers, and he is currently completing a lean industrial engineering book with Hiroyuki Hirano. He lives and works from his home in Tucson, Arizona.

Norman Bodek
Bodek@pcspress.com

Norman Bodek is the president of PCS Inc., a publishing and consulting company in Vancouver, Washington. In 1979, he started Productivity Inc. and Productivity Press and published hundreds of management books on productivity, quality, and lean manufacturing. At Productivity, he also developed and published numerous management training programs, training videos, ran major national conferences on productivity and quality, over 200 seminars and workshops a year, 50 study missions to Japan, and published five monthly newsletters. From his 65 trips to Japan, he met and published the works of Taiichi Ohno, and Dr. Shigeo Shingo, the co-creators of the Toyota Production System (lean manufacturing), and many other masters in manufacturing improvement. He even reprinted and republished the 1926 book of Henry Ford *Today and Tomorrow*. He introduced to the West important Lean tools and techniques discovered in Japan: JIT, Jidoka, 5S, Value Stream Mapping, SMED, QFD, CEDAC, Hoshin Kanri, Andon, Kanban, Poka-Yoke, etc.. He started the Shingo Prize and is a Shingo Prize winner.

Recently, Norman introduced Technicolor Corporation and other companies to Quick and Easy Kaizen. At Technicolor in Detroit, Michigan, prior to his workshops, the company received in 2001 only 250 suggestions with 113 implemented. In the last twelve months (ending October, 2005) the company received 26,000 suggestions with over 17,000 implemented, resulting in over $8,000,000 in cost savings. Norman is a consultant and frequent speaker on Kaizen, Kaikaku, and lean manufacturing. He is a frequent keynote speaker: APICS 2005 conference in Kansas City, IIE Lean conference in Los Angeles, Lean Accounting Summit in Dearborn, ISO 9000 in Las Vegas, and others. He has written over 100 published management articles and is the author of three prior books, The Idea Generator – Quick and Easy Kaizen (with Bunji Tozawai, *Kaikaku The Power and Magic of Lean* and *All You Gotta Do Is Ask* (with Chuck Yorke).

Recommend Reading

Akao, Yoji (ed). - *Quality Function Deployment: Integrating Customer Requirements into Product Design* – Productivity Press, 1990

———. *Hoshin Kanri: Policy Deployment for Successful TPM* – Productivity Press, 1991

Akiyama, Kaneo. - *Function Analysis: Systematic Improvement of Quality and Performance* – Productivity Press, 1991

Asaka, Tetsuichi and Kazuo Ozeki (eds.) - *Handbook of Quality Tools: The Japanese Approach* - Productivity Press, 1990

Ballis, John – *Managing Flow: Achieving Lean in the New Millennium to Win the Gold* – Brown Books, 2001

Barrett, Derm – *Fast Focus on TQM: A Concise Guide to Companywide Learning* - Productivity Press, 1994

Basadur, Min - *Simplex a Flight to Creativity* – Paperback Creative Education Foundation, 1995

Belohlav, James A. - *Championship Management: An Action Model for High Performance* - Productivity Press, 1990

Bodek, Norman – *The Idea Generator: Quick and Easy Kaizen – PCS Press, 2002*

Boyle, Daniel C. – Secrets of a Successful Employee Recognition System - Productivity Press, 1995

Camp, Robert C. - *Benchmarking: The Search for Industry Best Practices that Lead to Superior Performance* - Productivity Press, 1989

Campbell, John Dixon – *Uptime: Strategies for Excellence in Maintenance Management* - Productivity Press, 1995

Chowdhury, Subir - *Management 21C* - Pearson Professional Education, 2002

Christopher, William F. and Carl G. Thor - *Total Employee Involvement – Handbook for Productivity Measurement and Improvement* - Productivity Press, 1993

Cooper, Robin and Regine Stagmulder – *Target Costing and Value Engineering* – Productivity Press co-published with The

Institute of Management Accountant, 1997
Covey, Stephen R. Covey – *The 7 Habits of Highly Effective People* – A Fireside Book, Simon & Schuster, 1989

De Bono, Edward - *Lateral Thinking: Creativity Step-By-Step* – HarperCollins, 1990
———. *Six Thinking Hats* - Little Brown & Co, 1999 paperback
D'Egidio, Franco. - *The Service Era: Leadership in a Global Environment* - Productivity Press, 1990
Deming, W. Edward - *Out of the Crisis* - MIT Press, 1986
———. *Elementary Principles of Statistical Control of Quality* - Nippon Kagaku Gijutsu Remmei, Tokyo, Japan, 1950
———.and Mary Walton - *The Deming Management Methods* – Perigee, 1988
Dennis, Pascal - *Andy & Me: Crisis and Transformation On The Lean Journey* - Productivity Press - 2005
Dennis, Pascal – *Lean Production Simplified: A Plain Language Guide to the World's Most Powerful Production System* – Productivity Press - 2002
Fisher, Kimbell - *Leading Self-Directed Work Teams: A Guide to Developing New Team Leadership Skills* - McGraw-Hill Professional Publishing, 1999
Fitz-Enz, Jac - *The 8 Practices of Exceptional Companies: How Great Organizations Make the Most of Their Human Assets* – AMACOM, 1997
———.*The Roi of Human Capital: Measuring the Economic Value of Employee Performance* – AMACOM, 2000
———.and Jack J. Phillips - *A New Vision for Human Resources: Defining the Human Resources Function by Its Results* – Crisp Management Library, 1999
Ford, Henry. - *Today and Tomorrow* – Productivity Press, 1988 (Double Day 1926)
Fukuda, Ryuji – *Building Organizational Fitness: Management Methodology for Transformation and Strategic Advantage* – Productivity Press, 1994
———. *Managerial Engineering: Techniques for Improving Quality and Productivity in the Workplace* – Productivity

Press, 1983

———. *CEDAC: A Tool for Continuous Systematic Improvement* – Productivity Press, 1990

Galsworth, Gwendolyn D. - *Visual Systems: Harnessing the Power of a Visual Workplace* - American Management Association, 1997

Grief, Michelle. - *The Visual Factory: Building Participation through Shared Information* - Productivity Press, 1991

Gotoh, Fumio. - *Equipment Planning for TPM: Maintenance Prevention Design* – Productivity Press, 1991

Gross, Clifford M. – *The Right Fit: The Power of Ergonomics as a Competitive Strategy* - Productivity Press, 1991

Hall, Robert W. – *Attain Manufacturing Excellence* – McGraw-Hill Professional Publishing, 1988

Hartley, John R. – *Concurrent Engineering: Shortening Lead Times, Raising Quality, and Lowering Costs* - Productivity Press, 1992

Harrington H., James and Daryl Conner, Nicholas L. Horney – *Project Change Management : Applying Change Management to Improvement Projects* – McGraw-Hill Professional, 1999

———.and Erik K. C. Esseling (Contributor), Harm Van Nimwegen - *Business Process Improvement Workbook: Documentation, Analysis, Design, and Management of Business Process Improvement* - McGraw-Hill Professional, 1997

———.and Glen D. Hoffherr, Robert P. Reid - *The Creativity Toolkit: Provoking Creativity in Individuals and Organizations* – McGraw-Hill Professional,1998

Hatakeyama, Yoshio. - *Manager Revolution! A Guide to Survival in Today's Changing Workplace* - Productivity Press, 1986

Head, Christopher W. – *Beyond Corporate Transformation: A Whole System Approach to Creating and Sustaining High Performance* - Productivity Press, 1997

Hirano, Hiroyuki. - 5 *Pillars of the Visual Workplace: The Sourcebook for* 5S *Implementation* - Productivity Press, 1995

———. *JIT Factory Revolution: A Pictorial Guide to Factory Design of the Future* – Productivity Press. 1989

———*JIT Implementation Manual: The Complete Guide to Just-In-*

Time Manufacturing - Productivity Press, 1990
———. *5S for Operators: 5 Pillars of the Visual Workplace*

Productivity Press, 1996
Hobbs, Dennis P. - *Lean Manufacturing Implementation: A Complete Execution Manual for Any Size Manufacturer* - J. Ross Publishing, 2003
Horovitz, Jacques. - *Winning Ways: Achieving Zero-Defect Service* – Productivity Press, 1990
Hutchins, Greg - *ISO 9000: A Comprehensive Guide to Registration, Audit Guidelines, and Successful Certification* - John Wiley & Sons, 1997
Ichida, Takashi (Compiler) – *Product Design Review: A Method for error-free Product Development* - Productivity Press,1996
Ishiwata, Junichi. - *IE for the Shop Floor 1: Productivity Through Process Analysis* - Productivity Press, 1991
Ishikawa, Kaoru - *How to Operate QC Circle Activities*, JUSE, 1985
———. *QC Circle Koryo: General Principles of the QC Circle* – JUSE, 1980
———. *Quality Control Circles at Work: Cases from Japan's Manu facturing and Service Sectors* – Asian Productivity Organization, 1984
Imai, Masaaki - *Kaizen: The Key to Japan's Competitive Success* – McGraw Hill, 1986
Jackson, Thomas L. and Constance E. Dyer – *Corporate Diagnosis: Setting the Global Standard for Excellence* (Based on the work of Shigehiro Nakamura and Dr. Ryuji Fukuda) - Productivity Press, 1996
———. And Karen R. Jones – *Implementing a Lean Management System* - Productivity Press, 1996
Japan Institute of Plant Maintenance, eds.- *Autonomous Maintenance for Operators* - Productivity Press, 1997
———.TPM *Encyclopedia.* Atlanta: Japan Institute of Plant Maintenance – Productivity Press, 1996.
———. *Focused Equipment Improvement for TPM Teams* - Productivity Press, 1997
———.TPM *for Every Operator* - Productivity Press, 1996

———.TPM for Survivors - Productivity Press 1996

Japan Management Association - *Kanban and Just-In-Time at Toyota: Management Begins at the Workplace.* - Productivity Press, 1986

———. *Total Productivity: The New Science of TP Management* (Based on the work of Shigehiro Nakamura) – Productivity Europe, 1996

———. The *Canon Production System: Creative Involvement of the Total Workforce* – Productivity Press, *1987*

Japan Union of Scientist and Engineers (JUSE). - *TQC Solutions: The 14-Step Process* – Productivity Press, 1991

Johnson, H. Thomas - *Relevance Lost: The Rise and Fall of Management Accounting* - Harvard Business School Press, 1991

______ *Profit Beyond Measure : Extraordinary Results through Attention to Work and People*- Free Press, 2000

______ *Relevance Regained* – Free Press. 2002

Jones, Morgan D. - *The Thinker's Toolkit : Fourteen Powerful Techniques for Problem Solving* - Times Books, 1998

Juran, J. M, ed. - A *History of Managing for Quality: The Evolution, Trends, and Future Directions of Managing for Quality* - ASQC Quality Press, 1995

Kanatsu, Takashi. - *TQC for Accounting: A New Role in Companywide Improvement* – Productivity Press, 1991

Kaner, Sam and Lenny Lind, Catherine Toldi, Sarah Fisk, Duane Berger – *Facilitator's Guide to Participatory Decision-Making* - New Society Pub, 1996

Karatsu, Hajime. **-** *Tough Words For American Industry* - Productivity Press, 1988

———. *TQC Wisdom of Japan: Managing for Total Quality Control* - Productivity Press, 1988

Kato, Kenichiro. - *I.E. for the Shop Floor 2: Productivity Through Motion Study* - Productivity Press, 1991

Kaydos, Will. - *Measuring, Managing, and Maximizing Performance* - Productivity Press, 1991

Kirby, J. Philip and David Hughes – *Thoughtware: Change the Thinking and the Organization Will Change Itself* – Productivity Press, 1991

Kobayashi, Iwao - *20 Keys to Workplace Improvement*- Productivity

Press, 1990

Liebling, Henry E. – *Handbook for Personal Productivity* - Productivity Press, 1989

Lewis, C. Patrick – *Building a Shared Vision: A Leader's Guide to Aligning the Organization* - Productivity Press, 1996

Lu, David J. - *Inside Corporate Japan: The Art of Fumble-Free Management* - Productivity Press, 1987

Mauer, Rick – *Feedback Toolkit: 16 Tools for Better Communications in the Workplace* - Productivity Press, 1994

———. *Caught in the Middle: A Leadership Guide for Partnership in the Workplace* - Productivity Press, 1994

Maskell, Brian H. - *Performance Measurement for World Class Manufac turing: A Model for American Companies* - Productivity Press, 1991

———. *Software and the Agile Manufacturer: Computer Systems and the World Class Manufacturing* - Productivity Press, 1994

Merli, Giorgio - *Total Manufacturing Management: Production Organization for the 1990s* – Productivity Press, 1990

———. *Co-Makership: The New Strategy for Manufacturers* – Productivity Press, 1991

Miltenburg, John – *Manufacturing Strategy: How to Formulate and Implement a Winning Team* - Productivity Press, 1995

Mizuno, Shigeru (ed.). - *Management for Quality Improvement: The 7 New QC Tools* – Productivity Press, 1988

Ishiwata, Junichi – *IE for the Shopfloor 1: Productivity Through Process Analysis* - Productivity Press, 1991

Kaydos, Will – *Measuring, Managing, and Maximizing Performance* - Productivity Press, 1991

Kelley, Tom with Jonathan Littman – *The Art of Innovation* – A *Currency Book,* Doubleday, 2001

King, Bob - *The Idea Edge: Transforming Creative Thought into Organizational Excellence* - Goal/QPC, 1998

Laraia, Anthony C. and Patricia E. Moody, Robert W. Hall - *The Kaizen Blitz: Accelerating Breakthroughs in Productivity and Performance* – John Wiley & Sons, 1999

Latzko, William J. and David M. Saunders – *Four Days with Dr.*

Deming: A strategy for Modern Methods of Management – Addison-Wesley Publishing Company, 1995

Liker, Jeffrey, ed. - *Becoming Lean: Inside Stories of U.S. Manufacturers* - Productivity Press, 1997

Liker, Jeffrey, - *The Toyota Way*, - McGraw-Hill, 2004

Louis, Raymond J. - *Integrating Kanban with* MRPII: *Automating a Pull System for Enhanced JIT Inventory Management* – Productivity Press, 1997

Maskel, Brian H. - *Performance Measurement for World Class Manufacturing: A Model for American Companies* – Productivity Press, 1991

———. *Making the Numbers Count: The Management Accountant as Change Agent on the World Class Team* - Productivity Press, 1996

———. With LearnerFirst - *Putting Performance Measurement to Work: Building Focus and Sustaining Improvement* - Productivity Press, 1997

Maslow, Abraham, and Deborah C. Stephens, Gary Heil - *Maslow on Management* – John Wiley & Sons, 1998

Maurer, Rick. - *Feedback Toolkit: 16 Tools for Better Communication in the Workplace* - Productivity Press, 1994

McGregor, Douglas, and Warren G. Bennis – *Human Side of Enterprise: 25th Anniversary Printing* - McGraw-Hill Higher Education, 1985

Merrill, Peter – *Do It Right the Second Time: Benchmarking Best Practices in the Quality Change Process* – Productivity Press, 1997

Michalski, Walter J. - 40 *Tools for Cross-Functional Teams: Building Synergy for Breakthrough Creativity* - Productivity Press, 1998

———. *Tool Navigator: The Master Guide for Teams* – Productivity Press, 1997

———. *40 Top Tools for Manufacturers: A Guide for Implementing Powerful Improvement Activities* – Productivity Press, 1998

Mizuno, Shgeru (ed.) – *Management for Quality Improvement: The 7 New QC Tools* - Productivity Press, 1988

Monden, Yasuhiro. - *Toyota Production System: An Integrated Approach to Just-In-Time* - Norcross, GA: Engineering and Management Press, 1998

———. and Michiharu Sakurai (eds.). *Japanese Management Accounting: A World Class Approach to Profit Management* - Productivity Press, 1990

———. *The Toyota Management System: Linking the Seven Key Functional Areas* – Productivity Press, 1999

———. *Cost Reduction Systems: Target Costing and Kaizen Costing* - Productivity Press, 1995

———. *Japanese Cost Management* - World Scientific Pub Co;, 2000

———. *Cost Management in the New Manufacturing Age: Innovations in the Japanese Automobile Industry* – Productivity Press, 1992

Moriguti, Shigeiti (ed.) – *Software Excellence: A Total quality Management Guide* - Productivity Press, 1997

Nachi-Fujikoshi (ed.). - *Training for TPM: A Manufacturing Success Story* - Productivity Press, 1990

Nakajima, Seiichi. - *Introduction to TPM: Total Productive Maintenance* - Productivity Press, 1988

———.*TPM Development Program: Implementing Total Productive Maintenance* - Productivity Press, 1989

Moser-Wellman, Annette - *The Five Faces of Genius* – Viking Press 2001

Musashi, Miyamoto - *A Book of Five Rings* – (1584-1598) – The Classic Guide to Strategy – The Overlook Press, 1974

Nachi Fujikoshi Corporation. - *Training for TPM: a Manufacturing Success Story* - Productivity Press, 1990

Nagashima, Soichiro – *100 Management Charts* – Asian Productivity Organization, 1990

Nakamura, Shigehiro – *The New Standardization: Keystone of Continuous Improvement in Manufacturing* – Productivity Press, 1993

———. and Hideyuki Takahasi – *Go-Go Tools: Five Essential Activities for Leading Small Groups- Productivity Europe*

Nikkan Kogyo Shimbun, (eds.) - *Poka-Yoke: Improving Product Quality by Preventing Defects* - Productivity Press, 1989

———. *Factory Management Notebook Series: Mixed Model Production* - Productivity Press, 1991

———. *Factory Management Notebook Series: Visual Control Systems: Visual Control Systems* - Productivity Press, 1991

———. *Factory Management Notebook Series: Mixed Model Production* - Productivity Press, 1991

———. *Visual Control Systems - Factory Management Notebook Series:* Productivity Press, 1995

———. *TPM Case Studies* - Productivity Press, 1995

Northey, Patrick and Nigel Southway – *Cycle Time Management: The Fast Track to Time-Based Productivity Improvement* – Productivity Press, 1993

Ohno, Taiichi. - T*oyota Production System: Beyond Large-Scale Production* - Productivity Press, 1988

———.*Workplace Management* - Productivity Press, 1988

———. and Setsuo Mito - *Just-In-Time for Today and Tomorrow* – Productivity Press, 1988

Ono, Ken'ichi – *Visual Feedback: Making Your5S Implementation Click* - Productivity Press, 1996

Osada, Takashi – *The 5S's: Five Keys to a Total Quality Environment* – Asian Productivity Organization, 1991

Perigord, Michel. - *Achieving Total Quality Management: A Program for Action* – Productivity Press, 1991

Psarouthakis, John. - *Better Makes Us Best* - Productivity Press, 198

Puri, Subhash C. – *Stepping up to ISO 14000: Integrating Environmental Quality with ISO 9000 and TQM* – Productivity Press, 1996

———. *ISO 9000 Certification: Total Quality Management* – Productivity Press, 1995

Productivity Press Development Team – *Just-In-Time for Operators Learning Package*- Productivity Press,1998

Regan, Michael D. - *The Kaizen Revolution* - Holden Press, 2000

Robinson, Alan. - *Continuous Improvement in Operations: A Systematic Approach to Waste Reduction* - Productivity Press, 1991

———. And Same Stern - Corporate Creativity: How Innovation and Improvement Actually Happen - Berrett-Koehler, 1997

———.and Dean M. Schroeder - *Ideas are Free: How the Idea Revolution is Liberating People and Transforming Organizations* - Berrett-Koehler, 2004

Robinson, Charles J. and Andrew P. Cinder. - *Introduction to Implementing TPM: The North American Experience* - Productivity Press, 1995.

Robson, Ross E. - *The Quality And Productivity Equation: American Corporate Strategies For The 1990s* – Productivity Press, 1990

Sakurai, Michiharu – *Integrated Cost Management: A Companywide Prescription for Higher Profits and Lower Costs* - Productivity Press, 1995

Sarita Chawla and John Renesch – *Learning Organizations* - Productivity Press, 1995

Scanlan, Phillip M. – *The Dolphins Are Back: A Successful Quality Model for Healing the Environment* - Productivity Press, 1995

Scholtes, Peter R., Brian L. Joiner, and Barbara J. Streibel - *The Team Handbook, Second Edition* - Joiner, 1996

Schonberger, Richard J. – *Japanese Manufacturing Techniques: Nine Hidden Lessons in Simplicity* – Free Press, 1982

———.*The World Class Manufacturing: The Lessons of Simplicity Applied* – Free Press - 1986

———.*World Class Manufacturing: The Next Decade* – Free Press, 1996

Sekine, Keniche, Keisuke Arai, and Bruce Talbot. - *Kaizen for Quick Changeover: Going Beyond SMED* - Productivity Press, 1992

———.*One-Piece Flow: Cell Design for Transforming the Production Process* - Productivity Press, 1992

———.and Keisuke Arai.- TPM *for the Lean Factory: Innovative Methods and Worksheets for Equipment Management* – Productivity Press, 1998

Shetty, Y.K and Vernon M. Buehler (eds.). - *Competing Through Productivity and Quality* – Productivity Press, 1989

Shiba, Shoji, Alan Graham and David Walden – *A New American TQM: Four Practical Revolutions in Management* - Productivity Press, 1993

Shingo, Shigeo - *A Revolution in Manufacturing: The SMED System* – Productivity Press, 1983

———. *Zero Quality Control: Source Inspection and the Poka-*

yoke System – Productivity Press, 1986

———.. *The Sayings of Shigeo Shingo: Key Strategies for Plant Improvement,* - Productivity Press, 1987

———. *Zero Quality Control: Source Inspection and the Poka-Yoke System* - Productivity Press, 1986

———.A *Study of the Toyota Production System from an Industrial Engineering* Viewpoint - Productivity Press, 1989

———. *Quick Changeover for Operators* – Created by Productivity Press from Shingo's SMED System, 1996

———. *Mistake-Proofing for Operators: The ZQC System* - Created by Productivity Press from Shingo's SMED System, 1996

Shinohara, Isao (ed.) - *New Production System: JIT Crossing Industry Boundaries* – Productivity Press, 1988

Shirose, Kunio. - *TPM: New Implementation Program in Fabrication and Assembly Industries* - Japan Institute of Plant Maintenance

———.*TPM for Workshop Leaders* - Productivity Press, 1992

———.TPM *Team Guide* - Productivity Press, 1996

———. And Yoshifumi Kimura and Mitsugu Kaneda – *P-M Analysis: An Advanced Step in TPM Implementation* - Productivity Press, 1995

———. and Keisuke Arai *Design Team Revolution: How to Cut Lead Times in Half and Double Your Productivity* - Productivity Press, 1999

Steinbacher, Herbert R. and Norman L. Steinbacher – *TPM for America: What it is and Why You Need It* - Productivity Press, 1993

Suehiro, Kikuo – *Eliminating Minor Stoppages on Automated Lines* – Productivity Press, 1992

Sugiura, Tadashi and Yoshiaki Yamada – *The QC Storyline: A Guide to Solving Problems and Communicating the Results* – Asian Productivity Organization, 1995

Sugiyama, Tomo. - *The Improvement Book: Creating the Problem-Free Workplace* – Productivity Press, 1989

Suri, Rajan – *Quick Response Manufacturing: A Companywide Approach to Reducing Lean Times* - Productivity Press, 1998

Suzaki, Kiyoshi - *The New Manufacturing Challenge: Techniques for Continuous Improvement* - Free Press, 1987

———.*The New Shop Floor Management* - The Free Press, 1993.

Suzue, Toshio and Akira Kohdate. - *Variety Reduction Program (VRP): A Production Strategy for Product Diversification* –

Productivity Press, 1990

Suzuki, Tokutaro (ed.) – *TPM in Process Industries* - Productivity Press, 1994

Swartz, James B. – *The Hunters and the Hunted: A Non-Linear Solution for Reengineering the Workplace* - Productivity Press, 1994

Taguchi, Genichi, Subir Chowdbury and Shin Taguchi - *Robust Engineering: Learn How to Boost Quality While Reducing Costs & Time to Market* – McGraw - Hill Professional, 1999

Tajiri, Masaji and Fumioh Gotoh - *Autonomous Maintenance in Seven Steps: Implementing TPM on the Shop Floor* - Productivity Press, 1997

Takahasi, Yoshikazu andTakashi Osada – *TPM: Total Productive Maintenance* – Asian Productivity Organization, 1990

Tateisi, Kazuma. - *The Eternal Venture Spirit: An Executive's Practical Philosophy* – Productivity Press, 1989

Tel-A-Train and the Productivity Press Development Team – *The 5S System: Workplace Organization and Standardization* – 1998

Todorov, Branimir – ISO 9000 Required: Your Worldwide Passport to Customer Confidence - Productivity Press, 1996

Tozawa, Bunji – *Kaizen Teian 1: Developing Systems for Continuous Improvement Through Employee Suggestions* – Productivity Press, 1992

———. *Kaizen Teian 2: Developing Systems for Continuous Improvement Through Employee Suggestions* – Productivity Press, 1992

———.*The Idea Book: Improvement through TEI (Total Employee Involvement)* – Productivity Press, 1988

———. *The Improvement Engine: Creativity & Innovation Through Employee Involvement* - Productivity Press – 1995

———. *The Service Industry Idea Book: Employee Involvement in Retail and Office Improvement* – Productivity Press, 1990

Tsuchiya, Seiji – *Quality Maintenance: Zero Defects Through Equipment Management* - Productivity Press,

Uchimaru, Kiyoshi, Susumu Okamoto and Bunteru Kurahara – *TQP for Technical Groups: Total Quality Principles for*

Product Development - Productivity Press, 1993

Watson, Gregory H. - *Benchmarking Workbook: Adopting the Best Practices for Performance Improvement* - Productivity Press, 1992

Wellington, Patricia - *Kaizen Strategies for Customer Care: How to Create a Powerful Customer-Care Program—And Make It Work* –Financial Times Prentice Hall Publishing, 1996

Womack, James P. and Daniel T. Jones – *Lean Thinking, Banish Waste and Create Wealth in your Corporation* – Simon & Schuster, 1996

———. and Daniel T. Jones, and Daniel Roos. - *T*Al Jazeera*he Machine that*

Changed the World: Based on the Massachusetts Institute of Technology S-million dollar 5-year Study on the Future of the Automobile - Rawson Associates, 1990

Yasuda, Yuzo. - *40 Years, 20 Million Ideas: The Toyota Suggestion System* - Productivity Press, 1991

* The above list includes all of the books published by Norman Bodek at Productivity Press, most have his introduction. In addition Bodek and Productivity Press published 60 books in the Management Master Series each approximately 60 to 75 pages.

Testimonials

"Manufacturing is important."

Rebirth of American Industry is an important book. It blows the lid off many of the assumptions, processes, and metrics that have led our country down a road that has shipped millions of jobs overseas.

In order for the "Rebirth of American Industry" to take place, a great deal of pain will have to be incurred, unless we immediately challenge the existing norms and make some radical changes.

Death of waste, the elimination of all non-value added activities, must be pursued relentlessly. Birth of a new American heritage of producing with increased value to the customer, designing and delivering products, and providing services under world class conditions is mandatory.

In our free society the customer makes the choice of which products will be purchased and which will not. In a world of vast and instantly available information, with global competitors operating in our own backyards, the customer is indeed king. Companies that choose to ignore this fact do so at great peril.

Newsmagazines and commentators have blasted Wal-Mart, big box retailers, and global competition for our problems in manufacturing. The truth is that educated, knowledgeable customers with virtually unlimited choices wield more power than ever before.

Our failures in manufacturing are a result of failing to meet the needs of these customers. In the words of Walt Kelly's Pogo "We have met the enemy – and he is us."

Just as the customer has been empowered under our free economy, we must move forward in educating and empowering our manufacturing employees to better serve that customer.

We must align our strategies, business processes, and performance metrics to the voice of the customer, and aggressively go after our leaner, more agile competitors.

This book provides a prescription that should lead American companies to return manufacturing to our shores, just as many of our global competitors have chosen to join us here. - **Doug Nelson, CFPIM, CIRM, CSSBB, President, APICS, Portland Chapter**

"Before a rebirth is possible, the leaders of American industry have to wake up. The ideas in this book ring out clear and loud like a bell. This book is required reading for anyone who is committed to taking manufacturing into the future." **Jon Miller, President Gemba Research LLC**

This excellent book will make some enemies. It is outspoken, hard-hitting, and correct. The authors answer the question "why have so few American companies successfully transformed themselves into lean organizations." They take us back to the origins of lean at Ford Motor Company and Toyota, and contrast them with the modern American manufacturer.

The authors show that the management methods and accounting systems developed by Alfred Sloan and Pierre Dupont at General

Motors are fundamentally in opposition to lean thinking, yet these methods are ubiquitous and are even written into some of our laws and many of our accounting standards.

Companies that try to implement lean tools in production without dismantling and reinventing their management and accounting methods are doomed to failure. When senior managers focus on stock price and margins it is impossible for operations people to succeed with lean manufacturing beyond some production improvement.

The solutions advocated will be unpopular because they cut to the heart of "professional" management theory and show that the lean transformation must start not on the shop-floor but by active transformation in the executive offices. - **Brian H. Maskell, President of *BMA Inc.***, - author of *Performance Measurement for World Class Manufacturing* (1991), *Software and the Agile Manufacturer* (1993), *New Performance Measures* (1994), *Making the Numbers Count; the accountant as an agent of change* (1996), and the software product *Putting Performance Measurement to Work* (1999)

I enjoyed the book entirely. The overview of manufacturing is not known to all and I appreciate that review. I, as you know, have a background in Lean manufacturing and finance and the details were NOT known to me. And, it should be told more often. So I can only imagine how interesting it will be to people less informed about manufacturing history. This detail and thread that you have built from Ford, Toyoda, Sloan, and through to accounting systems will keep people interested. As an MBA'er, I am, and having a Lean background, I can fully relate to your position on the school of thought at the higher level of business education. I know that many CEO's and Presidents will agree with your encompassment of business structure from a different intuitive (common sense) point of view. Your book, shows that our intuition (common sense) is the product of our education. Therefore if the education is not correct then the intuition needs to change and this needs to be done through a change in our business education. I agree that we have been educated by a "slice" of thought, of someone's preference, and for some reason this is has been taught to us. I do believe that for MBAers we need an education that is not liberal arts nor technically specific; we need to establish a system that shows the flaws of 'systems and thought processes' so we can learn from other's mistakes. Currently, we learn from examples that are too simple and in their simplicity are correct, but as you have shown business is not a combination of simple things put together to create a whole but in fact it is a whole that can not be broken up into slices. This kind of approach, if I may be so bold, should be one where we learn from mentors and acquired wisdom rather than slices of knowledge.

On a less philosophical note, I have to let you know that towards the middle of this book I grabbed a pen, which I do not do often, and started underlying key ideas and counter intuitive statements, then I started circling entire sentences, then I grabbed a marker and highlighted more and more! I will take ideas from this book and incorporate them into my business and I know this book will influence our training packages. **Collin McLoughlin, President ENNA, Inc.**

In *Rebirth of American Industry*, William Waddell and Norman Bodek call American management onto the carpet. By showing how modern financial accounting (derived from practices from 8 decades past) drives American companies to non-lean measures. Mr. Waddell and Mr. Bodek have clearly demonstrated why American manufacturers continue to come up short when compared to their lean competitors. If unheeded, it could be the epitaph of a once-great manufacturing powerhouse. - **Bill Kluck, President, The Northwest Lean Networks**

The Idea Generator – Quick and Easy Kaizen

"The Idea Generator" brings continuous improvement down to Earth—and raises up to the heavens the importance of everybody documenting every implemented idea, however small. It's a message too few understand **—Richard J. Schonberger, President, Schonberger & Associates, Inc. and author of *Japanese Manufacturing Techniques: Nine Hidden Lessons in Simplicity***

"Lean systems will degrade without ongoing improvement from every employee through a myriad of simple, quick changes. What brings lean structures to life is people—people engaged in continuous improvement. Tozawa and Bodek provide deep insights into this fundamental ingredient of high performance companies."
—Dr. Jeffrey K. Liker, University of Michigan and author of Becoming Lean

Kaikaku — The Power and Magic of Lean

"Norman has always had fascinating stories about his travels around the world, the people he's met, and the things he's learned. Some of the most fascinating of these stories concern his friendship with Mr. Shigeo Shingo. As you read through Kaikaku, it is my sincere hope that you will appreciate the years of learning that went into its creation, as well as the great many things that Norman has brought to this country, and the world."
-Bill Kluck, President, NWLEAN.net

"One of the things you mentioned in your email to me was that you recommend I read Kaikaku. I purchased it a couple months ago and it just now made it to my reading rotation. If I had only known what a great book it is I would have devoured it the moment it arrived!" **—James A. Feltman, Sales Manager, Vorne Industries, Inc.**

"I cannot begin to explain how excited and energized I am around the theories you have both talked about at the APICS international conference and have written about in your book Kaikaku (which is written perfectly I must say....it's like talking to a friend and hearing their life stories as well)." **Nicole Yezierski, Yankee Candle**

All You Gotta Do Is Ask
By Chuck Yorke and Norman Bodek

Yes, All You Gotta Do Is Ask Your Employees to make their work easier, more interesting and to build their skills. The result is happier people, reduced costs, improved quality, improved safety, better customer service and faster throughput. Subaru in 2003 received 108.1 ideas per employee and saved $5,246 per employee.

"A simple approach and a very powerful message centering on the brain-power each employee can bring to work and how leaders can tap into it." – **David Veech, Manager Lean Certification Programs, University of Kentucky**

"From the Suggestion Box to the Toyota Production System--All You Gotta Do Is Ask has a lot of good ideas to save an organization time, effort, and money. It is well worth reading!" - **William C. Byham, Ph.D. , Chairman, DDI and author of Zapp!: The Lightning of Empowerment : How to Improve Productivity, Quality, and Employee Satisfaction**

All You Gotta Do Is Ask explains how to promote large numbers of ideas from your employees, something most organizations do poorly, if at all. The people who manage such organizations are usually either unaware of the power of employee ideas, or they don't know how to tap it. This easy-to-read book will show you why it is important to have a good idea system; how to set one up; and what it can do for you, your employees, and your organization. - Alan G. Robinson, professor University of Massachusetts and author of Ideas Are Free

Reviews of "Rebirth of American Industry"
by William H. Waddell and Norman Bodek

This excellent book will make some enemies. It is outspoken, hard-hitting, and correct. The authors answer the question "why have so few American companies successfully transformed themselves into lean organizations." They take us back to the origins of lean at Ford Motor Company and Toyota, and contrast them with the modern American manufacturer. The solutions advocated will be unpopular because they cut to the heart of "professional" management theory and show that the lean transformation must start not on the shop-floor but by active transformation in the executive offices. - **Brian H. Maskell, President of** *BMA Inc.*, **- author of *Putting Performance Measurement to Work***

Rebirth puts American management on the carpet; showing how modern accounting drives American companies to non-lean measures. It clearly demonstrates why American manufacturers continue to come up short when compared to their lean competitors. If unheeded, it could be the epitaph of a once-great manufacturing powerhouse. - **Bill Kluck, President, The Northwest Lean Networks**

I enjoyed the book entirely. This detail and thread that you have built from Ford, Toyoda, Sloan, and through to accounting systems will keep people interested. I will take ideas from this book and incorporate them into my business. - **Collin McLoughlin, ENNA, Inc**.

Rebirth of American Industry is an important book. It blows the lid off many of the assumptions, processes, and metrics that have led our country down a road that has shipped millions of jobs overseas. This book provides a prescription that should lead American companies to return to our shores just as many of our global competitors have chosen to join us here. - **Doug Nelson, President APICS, Portland Chapter**

"Before a rebirth is possible, the leaders of American industry have to wake up. The ideas in this book ring out clear and loud like a bell. This book is required reading for anyone who is committed to taking manufacturing into the future." - **Jon Miller, President, Gemba Research LLC**

PCS Press
809 SE 73rd Ave
Vancouver, WA 98664
360-737-1883 Fax
bodek@pcspress.com
http://www.pcspress.com

ISBN 0-9712436-3-8
90000
9 780971 243637